MAGNETOSTATISTICS

MAGNETOSTATISTICS

P.K. Aggarwal

CYBER TECH PUBLICATIONS
4264/3, Ansari Road, Daryaganj, New Delhi-110002 (India)
Ph.: 011-23244078. Fax: 011-23280028
E-Mail: cyberpublicationsdelhi@yahoo.com

MAGNETOSTATISTICS

P.K. Aggarwal

First Edition 2009

Published by :

G.S. Rawat for **Cyber Tech Publications**
4264/3, Ansari Road, Daryaganj, New Delhi-110002 (India)
Ph.: 011-23244078. Fax: 011-23280028
E-Mail: cyberpublicationsdelhi@yahoo.com

Printed at: Asian Offset Printers

PREFACE

This book *Magnestostatistics* language has been kept as simple as could be consistent with precision and brevity. for most of universities in India. The text have on purpose been kept fast, partly to reduce the volume of the book, and partly to allow for different approaches by the individual students. Further, the emphasis must be on the physical assumptions underlying mathematical procedures of integration, summation etc. The problems from an integral part of the text.

This book aim to provide the necessary foundation in wave optics which prepare the students for an intensive study of advanced topics at a later stags much of Nuclear Physics requires a good knowledge of physics. It is how the standard recognized practice in all advanced teaching methods the problems must not be limited to reproduction of some parts of the text, but must lead to projection of the students mind beyond the subject matter of the text, recent examination trends are also in that direction, therefore, we have not simply reproduced examination questions from different Universities but have built up a structure of problems which have a great teaching value in themselves.This book aim to provide the necessary foundation in wave optics which prepare the students for an intensive study of advanced topics at a later stags much of Magnestostatistics requires a good knowledge of physics.

—***Author***

Contents

1

Electromagnetic Sensitivity

INTRODUCTION

The sensitivity of a B.G. with only electromagnetic damping depends on the resistance of the circuit through which the damping current flows. For an instantaneous discharge, the sensitivity is given by,

$$\text{Charge sensitivity} = \left(\frac{2\pi}{T}\right) \times \text{current sensitivity}$$

$$\text{and voltage sensitivity} = \frac{\text{Current Sensitivity}}{\text{Resistance}}$$

The electromagnetic damping coefficient is $\left(\frac{m}{R}\right)$, say K when entire resistance of damping circuit is R and K_c its value when damping is critical corresponding to resistance R_c.

$$K = m/R$$

$$K = m/R_c$$

$$\frac{K_c}{K} = \frac{R}{R_c} = n, \quad R_c \text{ is the critical damping resistance}$$

and $$R = nR_c.$$

And as we know that the condition necessary for ballistic galvanometer is that T should be large and R should be large. Hence, increases the sensitivity.

Thus, we see that the resistance of the galvanometer and the period of vibration have great influence on the sensitivity of the galvanometer. Therefore, the sensitivity of a galvanometer is defined as the deflection in mm on a scale 1 metre away when a current of one micro ampere is passed when the resistance of the galvanometer is one ohm and the period of vibration one second.

Mega-Ohms Sensitivity

It is the number of mega-ohms to be put in series with the galvanometer coil so that a potential difference of one volt across the combination produces an deflection of one mm on a scale placed 1 metre away from the galvanometer mirror.

DAMPING

The oscillatory motion in the B.G. is given by,

$$\theta = Be^{-bt} \sin(\sqrt{K^2 - b^2}\, t + \beta)$$

However, the effort is made in the B.G. to make b as small as possible so that the motion is simple harmonic, but in actual practice b can never be zero so the amplitude of oscillation decreases with time.

The amplitude of motion is Be^{-bt} and

for t = 0, we have the amplitude $\theta_o = B$

$$\text{for } t = \frac{T}{4} = \frac{\pi}{2\sqrt{K^2 - b^2}} \text{ amplitude } \theta_1 = Be^{-\frac{b\pi}{2\sqrt{K^2-b^2}}}$$

$$\text{for } t = \frac{3T}{4} = \frac{3\pi}{2\sqrt{K^2 - b^2}} \text{ amplitude } \theta_2 = Be^{-\frac{3b\pi}{2\sqrt{K^2-b^2}}}$$

$$\text{for } t = \frac{5T}{4} = \frac{5\pi}{2\sqrt{K^2 - b^2}} \text{ amplitude } \theta_3 = Be^{-\frac{5b\pi}{2\sqrt{K^2-b^2}}}$$

$$\frac{\theta_1}{\theta_2} = \frac{\theta_2}{\theta_3} = \frac{\theta_3}{\theta_4} \cdots \cdots$$

$$= e^{+\frac{b\pi}{\sqrt{k_2 - b^2}}} = e\lambda = d.$$

The successive amplitudes θ_1, θ_2, θ_3 etc. are continuously decreasing and the ratio of one value to the next is always a constant d. This constant ratio d is called decrement and $\log_e d = \lambda$ is called the logarithmic decrement λ.

$$\text{And now, } \theta_1 = Be^{-\frac{b\pi}{2\sqrt{k_2-b^2}}}$$

or $$\theta_1 = Be^{-\lambda/2}$$

or $$\theta_1 = \theta_o e^{-\lambda/2}$$

Hence $$\frac{\theta_o}{\theta_1} = e^{-\lambda/2}$$

$$= \left[1 + \frac{\lambda}{2} + \frac{1}{1\times 2}\left(\frac{\lambda}{2}\right)^2 + \ldots\right]$$

$$\frac{\theta_o}{\theta_1} = 1 + \frac{\lambda}{2} \text{ (approximately)}$$

or $$\theta_o = \theta_1 \left(1 + \frac{\lambda}{2}\right)$$

Thus we can correct the effect of damping in B.G. although we cannot avoid it. Now the corrected charge,

$$q = \frac{c}{BA}\frac{T}{2\pi}\theta_1\left(1 + \frac{\lambda}{2}\right) \qquad \ldots(1)$$

Experimental Determination of the Logarithmic Decrement

We have established a relation,

$$\frac{\theta_1}{\theta_2} = \frac{\theta_2}{\theta_3} = \frac{\theta_3}{\theta_4} = \ldots = \frac{\theta_9}{\theta_{10}} = \frac{\theta_{10}}{\theta_{11}} = d = e\lambda$$

Multiplying all these factors, we get,

$$\frac{\theta_1}{\theta_{11}} = e^{10}\lambda$$

or $$\lambda = \frac{1}{10}\log_e \frac{\theta_1}{\theta_{11}} = \frac{2.3026}{10}\log_{10}\frac{\theta_1}{\theta_{11}}$$

Thus observing θ_1 and θ_{11} *i.e.*, first throw and eleventh throw, one can calculate λ.

Calibration of The B.G.

In equation (1.22), the charge q can be written as,

$$q = k\,\theta_1\left(1 + \frac{\lambda}{2}\right)$$

where k is ballistic constant given by,

$$k = \frac{c}{AB}\frac{T}{2\pi}.$$

And the current is given by,

$$i = \frac{c}{AB}\theta$$

or

$$\frac{c}{AB} = \frac{i}{\theta}.$$

$\frac{c}{AB}$ is called current reduction factor and is determined by sending a known steady current i_o through the galvanometer and observing the deflection θ. Then,

$$q = \frac{T}{2\pi} i_o \frac{\theta_1}{\theta}\left(1+\frac{\lambda}{2}\right) \qquad ...(1)$$

The time period T is determined with the stop watch. Hence K can be calculated.

HALL EFFECT

The fact that the charged particles in magnetic field are deflected has changed our basic concept that in the metal the current is carried only by electrons. If a conductor is placed in a magnetic field such that the direction of flow of current is at right angle to the magnetic field, a voltage is developed across the conductor in the direction perpendicular to both current and field. The phenomena by which voltage developed is called Hall effect and voltage itself is Hall voltage. A section of a metal bar carrying a current of current density j_x in the positive x-direction. The electrons are driven by a field $\vec{E}_x$ and drifting to the left. Since the current j_x is in the + x-direction, the current density j_x and the field $\vec{E}_x$ is related by

$$\vec{j}_x = \sigma \vec{E}_x$$

Now an external field $\vec{B}_z$ in the z-direction is applied. The deflecting force due to $\vec{B}_z$ acts on electrons. Consequently electrons are being deflected downward. But since they cannot escape at the bottom of the bar, they simply accumulate there until the surplus of negative charge at the bottom of the bar and the corresponding excess of positive charge at the top creates an electric corresponding excess of positive charge an electric field $\vec{E}_t$, due to this field the upward force eE_t exactly balances the downward force $ev_x B_z$ (due to B_z, since v_x and B_z are perpendicular).

In the steady state which is attained very quickly the average motion is again horizontal. Due to this field (E_t) a downward force acts on the positive ions. This force accelerates the bar downward.

The Lorentz force on an electron is,

$$\vec{F} = e(\vec{E} + \vec{v} \times \vec{B})$$

When the downward force balanced by up ward force of developed transverse field, the resultant force is zero.

$$F_y = 0 = e(E_y - v_x B_z)$$

or $$E_y = v_x B_z.$$

If d is width of the bar then the potential difference developed, across the bar,

$$V = E_y.d = v_x B_z d$$

or $$v_x = E_y / B_z$$

Now the current density,

$$j_x = Ne\, v_x$$

where N are number of charge carriers, *i.e.*, electrons per cubic meter. Substituting v_x, we get,

$$j_x = Ne. \frac{E_y}{B_z}$$

$$\frac{1}{Ne} = \frac{E_y}{j_x B_z} = R_H \qquad ...(1)$$

where R_H is called the Hall coefficient or constant and is negative for most metal where the current is carried by the free electrons.

The sign of Hall coefficient is positive for some metals like beryllium,. zinc and cadmium. The positive value of Hall coefficient predicts that the current is also carried by a positive charge. This fact is explained on the basis of band theory as an example in semiconductor the current is carried by both positive and negative. Eqn. (1) gives the number of current carriers per unit volumes.

THE FORCE ON A MAGNETIC DIPOLE IN A NON-UNIFORM EXTERNAL MAGNETIC FIELD

For simplicity, let us consider a small circular loop of radius r is placed in x-y plane. Further, consider an external field $\vec{B}$ of some other current system, such as solenoid, pointing along z-axis. The field is non-uniform. Its magnitude decreases as we proceed in the z-direction spacing of the field lines. Let us calculate the force acting on the magnetic dipole due to this external field. From the spreading out of the field lines, it is evident that there is an outward (radial) component of $\vec{B}$ everywhere. Let this radial component be $\vec{B}_r$ acting in outward direction as shown in the figure. Due to this field component, each element of the loop will experience a force.

$$d\vec{F} = Id\vec{l} \times \vec{B}_r$$

Its magnitude,
$$dF = IdlB_r$$

Because $\vec{B}_r$ is radial. The direction of this force is perpendicular to the plane of dl and B_r (or perpendicular to dl and $\vec{B}_r$) *i.e.*, – z-direction which is downwards if the direction of the current is as shown in diagram. Since we have considered a small loop, B_r can be taken to be constant for the entire loop. Therefore, the total downward force

$$F = \oint dF = IB_r \oint dl = I2\pi rB_r$$

or
$$\vec{F} = -2\pi r\ IB_r\hat{z} \qquad ...(1)$$

Let us consider a small cylinder of radius r and height dz situated close to the loop . Let us calculate the flux, passing through this cylinder. Since B is not uniform, but changing, its space rate of change in + z direction is $\frac{\delta B_z}{\delta z}$.

As we know, the flux of magnetic field is $B.d\vec{S}$. Therefore,

The outward flux from the curved surface of the cylinder

$$= 2\ \pi rdzB_r$$

The inwards flux from the lower end-face of the cylinder is

$$= \pi r^2 B_z$$

Since $\frac{\delta B_z}{\delta z}$ is the rate of change of B_z, the change for the height dz of the cylinder $= \frac{\delta B_z}{\delta z}\ dz$

Therefore, the field on the top face of the cylinder is

$$= B_z + \frac{\delta B_z}{\delta z}\ dz$$

Thus the outward flux from the top face of the cylinder is

$$= \pi r^2\left(B_z + \frac{\delta B_z}{\delta z} dz\right)$$

Therefore, the net outward flux,

$$= 2\,\pi r dz B_r + \pi r^2 \left(B_z + \frac{\delta B_z}{\delta z} dz\right) - \pi r^2 B_z$$

$$= 2\pi r dz B_r + \pi r^2 \frac{\delta B_z}{\delta z} dz$$

But we know that div $\vec{B} = 0$, which means that the net flux of the magnetic field $\vec{B}$ out of any given volume must be zero, hence.

$$= 2\pi r dz B_r + \pi r^2 \frac{\delta B_z}{\delta z} dz = 0$$

or $$B_r = -\frac{r}{2} \frac{\delta B_z}{\delta z}$$

The negative singe indicates the rate of decrease of B_r with increasing z. Substituting in equation (1), the force on the magnetic dipole is given by,

$$F = 2\pi r I \frac{r}{2} \frac{\delta B_z}{\delta z}$$

$$= I\pi r^2 \frac{\delta B_z}{\delta z}$$

But $\pi r^2 = A$ area of the loop

and $IA = m$, the magnetic dipole moment,

therefore, $$F = m \frac{\delta B_z}{\delta z} \quad ...(2)$$

The magnetic dipole moment points in the positive z-direction and in our case parallel to the direction of the external magnetic field. We thus conclude that when the moment of the dipole is parallel to the external field, the force acts in direction of the increasing field strength.

Clearly, from relation (2) the force is independent of the space of the loop, therefore, small loops of all shapes will experience the same force in the same magnetic field. If the magnetic moment $\vec{m}$ is not in the direction of the field and making certain angle with $\vec{B}$and the components of $\vec{B}$varying in different directions say, x, y, and z, then,

$$\left.\begin{aligned}\vec{F}_x &= \vec{m}.\text{grad } B_x\\ \vec{F}_y &= \vec{m}.\text{grad } B_y\\ \vec{F}_z &= \vec{m}.\text{grad } B_z\end{aligned}\right\} \qquad ...(3)$$

when = 0, then in equation (2) force is also zero.

THE DEFINITION OF $\vec{B}$

We have defined electric field $\vec{E}$ in terms of force experienced by a test charge q_o, which depends on the situation of the charge q_o. The fact that the force on an electric charge depends not only on its situation, but also on the velocity of the charge is considered for defining $\vec{B}$. A similar approach would be to define the field $\vec{B}$ of a moving charge in terms of force experienced by it.

Let us bring a positive test charge with velocity v at a point P situated in magnetic field. The test charge experiences a deflecting force. The magnetic induction or the magnetic field vector $\vec{B}$ is defined in terms of this force and other measurable quantities. This magnetic force has a strange directional character. At any particular point in space, both the direction of the force and its magnitude depend on the direction of the motion of the particle. At every instant the force is always at right angles to the velocity vector, also at any particular point, the force is always at right angles to a fixed direction in space, and finally the magnitude of the force is proportional to the component of the velocity at right angles to this unique direction. It is possible to describe all these behaviour by defining the magnetic field vector $\vec{B}$ as follows.

If we vary the direction of $\vec{v}$ through point P, keeping the magnitude of $\vec{v}$ unchanged, though $\vec{F}$ will always remain at right angles to $\vec{v}$, its magnitude F changes and for particular orientation the force becomes zero. We define this direction as the direction of $\vec{B}$.

Now, if the test charge is again oriented such that it moves at right angles to this direction of $\vec{B}$ for which force is zero, then the magnitude of force is maximum, say this force is denoted by $F\perp$ and we define the magnitude of $\vec{B}$ from the measured magnitude of this maximum force $F\perp$, *i.e.*,

$$B = \frac{F\perp}{q_o v} \qquad ...(1)$$

or in general,

$$\vec{F} = q_o \vec{v} \times \vec{B} \quad ...(2)$$

This equation is known as the Lorentz force law. Because for relation (1.1), $\vec{B}$ and $\vec{v}$ are right angles, the magnitude of deflecting force,

$$F = qo\ B\ \text{Sin}\ \theta \quad ...(3)$$

$\vec{F}$ is right angles to the plane formed by $\vec{v}$ and $\vec{B}$, will always be right angles to $\vec{v}$ and also $\vec{B}$ and thus always be a side ways force.

Unit

The RMKS unit of $\vec{B}$from equation (2) is

$$= \frac{\text{Newtons}}{\text{Coul} - \text{metre / sec}} = \frac{\text{Newtons}}{\text{ampere.metre}}$$

Since, coul/sec. = ampere.

or as will be shown later, the weber per square metre.

Work Done in Moving a Test Charge in a Static Magnetic Field

Since force is always right angles to the direction of motion, for an element of the path $d\vec{L}$, the work done is $\vec{F}.d\vec{L}$. This work done is zero because $\vec{F}$ and $d\vec{L}$ are right angles. *Thus, a static magnetic field cannot change the kinetic energy of a moving charge.* It can only deflect it sideways, whereas when a charge moves in static electric field, it kinetic energy changes and is given by

$$\left(\frac{1}{2}mv^2 = eV\right), \text{ where V is potential.}$$

THE LORENTZ FORCE

If a charged particle moves through a region where both an electric and a magnetic field are present, then the resultant force is given by

$$\vec{F} = q_o\vec{E} + q_o\vec{v} \times \vec{B}$$

$$= q_o[\vec{E} + \vec{v} \times \vec{B}] \quad ...(1)$$

This resultant force is called Lorentz force. This equation defines the field $\vec{E}$and $\vec{B}$ in free space, where the force on a moving test charge can be measured.

FORCE ON A CURRENT CARRYING WIRE

The equation (1) which gives the force on moving charge in a magnetic field, can be used for calculating the force exerted by a magnetic field on a wire carrying current. Basically the current i in a metal wire is carried by the free or conduction electrons.

If there are N charges per unit volume so number of charges in volume dV = NdV, therefore the force $d\vec{F}$ due to these changes is given by

$$d\vec{F} = NdV\,(q\vec{v} \times \vec{B})$$

where $(q\vec{v} \times \vec{B})$ is force on the charge q, but

Nqv = J, the current density,

Hence,

$$d\vec{F} = (\vec{J} \times \vec{B})\,dV$$

Let us apply it to the case of a cylindrical wire of cross-section A. Then we get force dF due to length dl.

$$d\vec{F} = (\vec{J} \times \vec{B})\,Adl,$$

$$[\text{since } dV = Adl]$$

$$= \vec{J}\,Adl \times \vec{B}$$

But, J = i/A

hence, $$d\vec{F} = id\vec{l} \times \vec{B} \qquad ...(2)$$

Therefore, for the total wire of length l, the resultant force is obtained by integrating the above relation, *i.e.*,

$$\vec{F} = \int d\vec{F} = i\vec{l} \times \vec{B} \qquad ...(3)$$

idl is called " current element". From Eq. (2), the force $d\vec{F}$ is perpendicular to $\vec{B}$ and $id\vec{l}$. The relation (3) shows that force depends on the total current and not on the amount of charge carried by each particle or even its sign. The direction of $\vec{B}$ is determined by means of the right-hand rule for linear currents. If the wire is grasped in the palm of the right hand such that thumb pointing in the direction of the current then the fingers encircling the wire point in the direction of the magnetic field or lines of Magnetic flux. This is applicable for linear conductors. The right-hand rule for circular currents states that if the direction of the circular current coincides with the direction of the closed fingers, then the thumb will point in the direction of the magnetic field or the lines of the magnetic flux.

THE FORCE PER CURRENT ELEMENT OR MAGNETIC FLUX DENSITY $\vec{B}$

The magnetic induction (or magnetic field vector) $\vec{B}$ is also called the magnetic flux density $\vec{B}$. The quantity idl = current × length = current moment. Unit of the current moment is ampere-metre. In electrostatics the electric field intensity $\vec{E}$ is defined as the force per unit ;charge. In magnetostatics, the magnetic flux density $\vec{B}$ is defined as the force per current moment .. The magnitude B from eq. (1) is,

$$B = \frac{F}{il} = \frac{\text{Force}}{\text{Current moment}} \quad ...(1)$$

The unit for B is the Newton/ampere-metre, or as will be shown later, the weber per metre^2. In electrostatics we define $\vec{E}$ as follows

$$\vec{E} = \frac{\vec{F}}{q} = \frac{\text{Force}}{\text{Charge}} \quad ...(2)$$

In order to establish an analogy with electrostatics, *the product il is called pole strength denoted by qm, them* equation (1) reduces the

$$\vec{B} = \vec{F}/q_m \quad ...(3)$$

we define $\vec{B}$ as the force per pole.

Because of the analogy between (2) and (3) $\vec{B}$ *is appropriately called the magnetic field intensity.* $\vec{B}$ is also called as the *magnetic flux density or magnetic induction.* The equation (1) is used to define $\vec{B}$ and is derived from the basic definition of $\vec{B}$ given by equation (2). But the definition given her is more clear and a good analogy with $\vec{E}$ is established as shown in Eq. (3).

MAGNETIC FLUX ϕ_B

The magnetic flux denoted by ϕ_B can be defined in exact analogy with electric flux ϕ_E or in general flux of any vector field discussed earlier. The net ourward flow or flux is the average outdrawn normal component of the field $\vec{B}$ times the area of the surface.

ϕ_B = (average normal component of $\vec{B}$) × area

The magnetic flux through area dS as shown in Fig. 1.4

$$d\phi_B = B\ dS\cos\theta$$
$$= \vec{B}\ .\ \hat{n}dS$$

where θ is the angle between $\vec{B}$ and normal to the area dS and $\hat{n}$ is unit vector along this normal. In general for any surface we have

$$_B = \int_S \vec{B}.d\vec{S} \qquad ...(1)$$

where integral is taken over the entire surface.

Unit

The rationalised MKS unit for magnetic flux is obtained from equation (1),

$$\text{Unit of } \phi_B = \frac{\text{Newton}}{\text{Amp. metre}}\ \text{metre}^2$$

Since unit of B is Newton/ampere-metre and Area = metre^2, therefore, unit of

$$\phi_B = \frac{\text{Newton}}{\text{Ampere}}$$
$$= \text{Weber}$$

Thus, in place of Newton-metre/ ampere, a new unit weber is introduced. Also,

$$\text{weber} = \frac{\text{Newton.metre}}{\text{Coul / sec}} = \text{Volt.sec.}$$

$$\text{since Volt} = \frac{\text{Newton.metre}}{\text{Coulomb}}$$

For the magnetic field $\vec{B}$, we have, earlier,

$$\frac{\text{Newton}}{\text{amp. metre}} = \frac{\text{Newton.metre}}{\text{ampere. metre}^2}$$

$$\text{But, 1 Weber} = \frac{\text{1 Newton.metre}}{\text{1 ampere}}$$

hence, unit for $\vec{B}$ $= \dfrac{\text{Weber}}{\text{metre}^2}$. ...(2)

We can obtain directly the same result from Eq. (2), The $\vec{B}$ is called flux density and defined as the number of magnetic flux lines passing per unit area through a surface normal to the lines. Hence, the total flux is the measure of number of lines passing over the area through a surface normal to the lines.

APPLICATIONS OF B.G.

The B.G. is used for various measurements. Some of very important are mentioned below :

1. Comparison of the emfs of two cells
2. Comparison of two capacitances
3. Determination of the following:
   (a) Angle of dip and components of earth magnetic field
   (b) Self-inductance
   (c) Mutual inductance
   (d) Magnetic susceptibility and permeability
4. Measurement of:
   (a) Strong magnetic field
   (b) High resistance.

Comparison of Emfs of Two Cells

A suitable condenser C is first charged with cell of emf ξ_1 and then discharge through the BG and observed the throw say it is θ_1. Similarly charge capacitance with ξ_2 and observed the throw say θ_2. Then,

$$q_1 = c\xi_1 = K\theta_1\left(1+\frac{\lambda}{2}\right)$$

and $$q_2 = c\xi_2 = K\theta_2\left(1+\frac{\lambda}{2}\right)$$

on dividing

$$\frac{\xi_1}{\xi_2} = \frac{\theta_1}{\theta_2} \quad ...(1)$$

The charge and discharge are easily effected with the help of charge discharge key K.

Comparison of Two Capacitances

Charge the two condensers with same battery and discharge through the BG and observed the deflection in both cases and say θ_1 and θ_2 successively. Then,

$$q_1 = c_1\xi = K\theta_1\left(1+\frac{\lambda}{2}\right)$$

$$\text{and } q_2 = c_2\xi = K\theta_2\left(1+\frac{\lambda}{2}\right)$$

on dividing

$$\frac{c_1}{c_2} = \frac{\theta_1}{\theta_2} \quad ...(2)$$

Ballistic Constant

If we use a standard capacitance c and charge it to a known emf ξ and then discharge it through the B.G. when the throw q is observed, then the B.C. constant K is easily calculated.

$$q = \xi c = K\theta\left(1+\frac{\lambda}{2}\right)$$

or

$$K = \frac{\xi c}{\theta\left(1+\frac{\lambda}{2}\right)} \quad ...(3)$$

DIRECT CURRENT AMMETER AND VOLTMETER

(or Conversion of galvanometer to Ammeter and Voltmeter)

A direct current ammeter or voltmeter operates on the principle that a torque is exerted on a coil carrying a current in a magnetic field. As in the ballistic galvanometer, a magnetic field is provided by curved pole pieces. Consequently the magnetic torque = NiAB,

where, B = Magnetic field

A = Area of the coil

i = Current through the coil

N = Number of the coils

Due to this torque the coil rotates. The rotation of the coil is opposed and the restoring torque acts due to spring by which coil is pivoted. When an ammeter or voltmeter is used in a circuit, there is a steady current in the meter coil. Equilibrium is established when the coil is at rest in a deflected position. Then,

$$NiAB = c\theta \qquad ...(1)$$

where c is constant restoring force per unit twist for a given coil. Clearly, the angle θ of deflection is proportional to the current in the coil as NAB are constant for a given ammeter, *i.e.*,

$$\theta \propto i$$

Construction of Ammeter

The internal circuit of a d.c. ammeter. Applying the Kirchhoff's laws to the closed loop we get,

$$i_g R_g - i_s R_s = 0 \text{ (from voltage law)} \qquad ...(2)$$

where the subscript g refers to the coil or galvanometer and s to the shunt.

From the Kirchhoff's current law, we get,

$$i - i_g i_s = 0$$

or

$$i = i_g + i_s \qquad ...(3)$$

where i is the current measured by the ammeter. After eliminating i_s from the above two equations, we have,

$$i = i_g \left(1 + \frac{R_g}{R_s}\right) \qquad ...(4)$$

Clearly, i is proportional to i_g and, therefore, it is possible to calibrate the ammeter to read i directly. The range of ammeter depends on the maximum value of the current (i_g) which the coil can carry safely and on the values of R_g and R_s *Thus an ammeter is just a galvanometer with very low resistance,* this is necessary to that ammeter may not appreciably alter the current in the circuit which is to be measured. Therefore a shunt resistance is connected in parallel.

MAGNETIC FLUX OVER A CLOSED SURFACE (OR GAUSS' LAW IN MAGNETOSTATICS)

The flux tubes (or field lines) or a static electric field originate and end on electric charges. On the other hand, tubes of magnetic flux (or lines of magnetic field $\vec{B}$) are continuous, that is they have no sources or sinks. This is fundamental difference between static electric and magnetic fields. To describe this continuous nature of lines of $\vec{B}$ or flux tubes, it is said that magnetic field $\vec{B}$ is solenoidal (as we have discussed in chapter of vectors that a field is solenoidal when the lines are closed curves and do not have sources or sinks). Since magnetic field $\vec{B}$ is continuous, the magnetic flux entering any region is equal to the flux leaving it. Hence, when integration in equation (1) is carried out over a closed surface, the result must be zero, or

$$\oint_S \vec{B}.d\vec{S} = 0.$$

This is the Gauss's law applied to the magnetic field. Since flux entering any volume equals to flux leaving, the net flux over the volume is zero. Mathematically, it means at any point,

$$\text{div } \vec{B} = 0$$

$$\nabla . \vec{B} = 0$$

TORQUE ON A CURRENT COIL IN UNIFORM MAGNETIC FIELD

A coil carrying a current in magnetic field experiences a torque. This is the basic principle underlying the operation of direct current measuring instruments, such as ammeters, voltmeters and galvanometers. We shall calculate the torque exerted on the one turn rectangular coil DEFG suspended in a uniform field $\vec{B}$. The direction of $\vec{B}$ being parallel to the plane of the coil, the magnitude of force on the side DE is according to equation (1),

$$F_{DE} = ih\ B. \text{ (angle between B and ih is 90)}$$

The direction of this force is normal to the page toward the reader. Similarly, the magnitude of force on FG is,

$$F_{FG} = ih\ B$$

and the direction of F_{FG} is normal to the page away from the reader because the direction of current is just opposite in De and FG arms. The *force F_{DE} and F_{FG} are equal in magnitude and opposite in direction and, therefore, constitute a couple.* The torque τ exerted by this couple on the coil,

$$\tau = ihl\ B,$$

since Torque = one of the forces × perpendicular distance.

There is no force on the side FE since a current element EF makes an angle of zero degrees with B or in other words EF and $\vec{B}$ are parallel, hence the product $(i\vec{l} \times \vec{B})$ is zero. Similarly, the force on the side GD is zero since the current here makes an angle of 180 degrees with $\vec{B}$. Thus, the total torque on the coil is,

$$\tau = ihl\,B$$

or $$\tau = i\,AB$$

where A is the area of the coil = lh.

If the coil has N turns instead of one turn, the torque is given by,

$$\tau = NiAB.$$

Finally, if the normal to the plane of the coil makes an angle θ with $\vec{B}$, as,

$$\tau = NiAB\sin\theta \qquad ...(2)$$

whereas electric dipole when placed in an external electric field it experiences a torque given by,

$$\tau = \vec{P} \times \vec{E}$$

$$\tau = PE\sin\theta$$

where $\vec{P}$ is the electric dipole moment.

Let us compare this equation with the expression for torque on a current loop, *i.e.*,

$$= (NiA)\,B\sin\theta$$

In each case the field ($\vec{E}$ or $\vec{B}$) appears. *Comparison suggests that (NiA) in equation (2) can be taken as magnetic dipole moment m, analogous to* $\vec{P}$. Hence,

$$\vec{m} = NiA \qquad ...(3)$$

then, $$\tau = \vec{m} \times \vec{B} \qquad ...(4)$$

Thus, coil behaves as a magnetic dipole. The direction of magnetic dipole moment lies along the axis of loop by the following rule. Let the fingers of the right hand curl around the loop in the direction of the current, the extended right hand thumb will then point in the direction of $\vec{m}$.

Unit

The unit of magnetic dipole moment from equation (3) is ampere. Metre^2.

Potential Energy of a Current Loop

Since a torque acts on a current loop when placed in an external field, it follows that work (positive or negative) must be done by an external agent to change the orientation of such dipole. This work done is restored as potential energy. This energy may be taken to be zero for any arbitrary position, taking that potential energy be zero when $\vec{m}$ and are at right angles that is, when $\theta = 90°$.

The magnetic potential energy is equal to the work done by the external agent in turning the loop from zero potential energy to a position θ. Thus,

$$U = \int_{90}^{\theta} \tau \, d\theta = \int_{90}^{\theta} NiAB \sin\theta \, d\theta$$

$$= mB \int_{90}^{\theta} \text{Sin}\,\theta \, d\theta$$

$$= -\,mB\ \text{Cos}\ \theta$$

or $$U = -\,\vec{m}.\vec{B} \qquad ...(5)$$

which is in perfect correspondence with potential energy of electric dipole,

$$U = -\ \vec{P}.\vec{E}$$

EQUIVALENCE OF CURRENT LOOP TO A SHORT MAGNETISED BAR

Let us consider a bar magnet of pole strength Q_m placed in magnetic field B making an angle θ with the axis of magnet.

Torque = Force × perpendicular distance

$$= Q_m Bl \sin\theta$$

$$= \vec{m} \times \vec{B}$$

where m = magnetic moment of bar, magnet given by

m = pole strength × distance between the poles.

Comparing this result with that of a torque due to a single coil carrying current, they are identical. So that we can say that a current carrying coil is equivalent to a bar magnet, its magnetic dipole moment must be iA. The pole strength in magnet is equivalent to current moment in coil.

However, the equivalence is valid for coil of any shape. Therefore, the equation (1) is valid even if the coil is not rectangular. Let us consider a coil. The plane coil may be divided into a large number of narrow rectangular circuits. The current in the coil (or loop) remains same because in the common boundary of any two rectangular circuits, there are two equal but oppositely directed currents. The torque due to the j-th rectangular circuit is,

$$\tau_j = iA_j\ B\ \text{Sin}\ \theta$$

Total torque τ is given by,

$$\tau = \sum_{J=1}^{J=n} \tau j = iB\ \text{Sin}\ \theta \sum_{J=1}^{J=n} Aj$$

$$= iAB\ \text{Sin}\ \theta \qquad ...(1)$$

where, $A \sum_{J=1}^{J=n} Aj$ is the total area of the coil. This is, of course, the same result as expressed in equation (1).

Thus, any current carrying coil is equivalent to a magnet (or magnetic dipole).

BALLISTIC GALVANOMETER

The ordinary galvanometers measure current. The B.G. measures charge. The B.G. are of two types, namely,

(i) The moving magnet B.G.

(ii) The moving coil B.G.

The Moving Coil Ballistic Galvanometer

A ballistic galvanometer consists essentially of a rectangular coil suspended in a magnetic field as shown in Fig. 1.9. Let the current i flow through the coil. There is no resultant force along bc and ad which are parallel to $\vec{B}$. There is equal to opposite force on ab and dc. The torque is given by,

$$\tau = iAB$$

where A is area of coil ib.

If the current i acts for a short interval dt, then the moment of impulse is iABdt and total moment of impulse,

$$\int_0^T iAB\,dt = qAB, \text{ since } q = \int_0^T i\,dt$$

But moment of impulse = mass × velocity × distance from the axis of rotation

= Momentum × distance for the axis of rotation

= angular momentum.

If I is moment of inertia of the coil about the suspension fibre and ω is angular velocity, then angular momentum = Iω

$$q\,AB = I\omega \qquad ...(1)$$

The kinetic energy of the system is $I\omega^2/2$ and is used up in twisting the suspension fibre. If c is the restoring force per unit twist, the couple on twist θ is cθ and work done for an additional twist dθ is cθ dθ so that the whole work done in twisting the suspension fibre is,

$$\int_0^\theta c\theta\,d\theta = \frac{1}{2}c\theta^2$$

therefore,

$$\frac{1}{2}c\theta^2 = \frac{1}{2}I\omega^2$$

or

$$\omega^2 = \frac{c}{I}\theta^2$$

$$= \frac{q^2A^2B^2}{I^2} \qquad \text{from equation (1)}$$

or

$$\frac{c}{I}\theta^2 = \frac{q^2A^2B^2}{I^2}$$

or

$$q^2 = \frac{c^2}{B^2A^2}\,\frac{I}{c}\theta^2$$

Now the period of vibration of a body of moment of inertia I, when c is the restoring couple per unit twist is given by,

$$T = 2\pi\sqrt{\frac{I}{c}}$$

or $$\frac{I}{c} = \frac{T^2}{4\pi^2}$$

therefore, $$q^2 = \frac{c^2}{B^2A^2}\frac{T^2}{4\pi^2}\theta^2$$

or $$q = \frac{c}{BA}\frac{T}{2\pi}\theta \quad ..(2)$$

or $$q = K\frac{T}{2\pi}\theta \quad ...(3)$$

where $$K = c/BA.$$

This gives a relation between the charge flowing and the ballistic throw θ of the galvanometer, where K is the galvanometer constant. The torque or couple on suspension fibre is cθ. Therefore,

$$iAB = c\theta$$

$$i = \frac{c}{BA}\theta. \quad ...(4)$$

Current Sensitivity

Quantitatively, the current sensitivity of a galvanometer is defined as the deflection, as read from the scale per unit current. So from equation (1),

$$\text{current sensitivity} = \frac{\text{deflection}}{\text{current}}$$

$$= \frac{\theta}{i} = \frac{BA}{c} \quad ...(5)$$

As we know that mirror and scale arrangement is used for reading the deflection, if the current i is in amp., the scale deflection θ in millimetres and scale distance L in metres, the current sensitivity,

$$Si = \theta/Li$$

If i = 1 amp, then S_i is a large number for a given value of L. Consequently for convenience we prefer microampere which gives microampere current sensitivity.

Micro-Ampere Sensitivity

is the deflection in mm produced on a scale 1000 mm away from the galvanometer mirror, when a current of one micro ampere (10^{-6} amp) is passed in the galvanometer.

Charge Sensitivity or Quantity Sensitivity

The charge sensitivity of the B.G. is defined as the deflection per unit charge. So from equation (3),

$$\text{Charge sensitivity} = \frac{\text{deflection}}{\text{charge}}$$

$$= \frac{\theta}{q} = \frac{2\pi}{T}\frac{BA}{c}$$

But from equation (5) $\frac{BA}{c}$ is current sensitivity

$$\text{Hence charge sensitivity} = \frac{2\pi}{T} \times \text{current sensitivity.} \qquad ...(6)$$

Hence the charge sensitivity of a B.G. is $\frac{2\pi}{T}$ times the current sensitivity.

Voltage Sensitivity

The voltage sensitivity under any given conditions is the deflection per unit voltage and is consequently, current sensitivity divided by the resistance, *i.e.*,

$$\text{Voltage sensitivity} = \frac{\text{Current Sensitivity}}{\text{Resistance}}$$

The resistance is that of entire circuit and not that of the instrument alone. Voltage sensitivity is large number, hence for convenience, we prefer micro-volt sensitivity.

Micro-volt Sensitivity

Micro-volt sensitivity is the deflection in millimetre on a scale 1 metre away from the galvanometer mirror, when potential difference of one micro-volt (10^{-6} volts) is maintained across the galvanometer terminals.

Figure of Merit of Galvanometer

Figure of merit of galvanometer is the current which will produce a deflection of one scale division, when we use lamp and scale arrangement, it is the current which will produce a deflection of 1 mm on a metre scale one metre away from the galvanometer mirror.

It may be noted here that the current sensitivity or charge sensitivity is sometimes defined as the current or charge per unit deflection and then it is reciprocal of what have been defined here by equations (5) and (6). We prefer to call it figure of merit of the galvanometer,

$$\text{Figure of merit} = \frac{q}{\theta} \text{ or } \frac{i}{\theta}.$$

CONDITION WHEN A MOVING COIL GALVANOMETER IS DEAD BEAT OR BALLISTIC

In deriving equation (1) we have assumed that whole of the kinetic energy of the coil is used up in twisting the fibre and neglected the retarding forces present namely, (1) due to friction of air, (2) viscosity of suspension fibre and (3) induced current in any neighbouring mass or metal. The first two of these factors constitute mechanical damping, while the third factor accounts for electromagnetic damping. The retarding couple due to damping varies as the angular velocity = $d\theta/dt$, where p is coefficient of mechanical damping.

When the circuit is closed, we have in addition the induced current produced in the coil which varies as the rate of change of displacement and inversely as the total resistance of the circuit, hence the retarding couple is $\frac{m}{R}\frac{d\theta}{dt}$, where the constant m involves the magnetic flux due to the magnet and the area of coil. Thus, the total retarding couple due to damping

$$= \left(p + \frac{m}{R}\right)\frac{d\theta}{dt}.$$

Now the angular acceleration is $\frac{d^2\theta}{dt^2}$.

Hence, according to Newton's second law, equation of motion is,

$$\frac{Id^2\theta}{dt^2} = -c\theta - \left(p + \frac{m}{R}\right)\frac{d\theta}{dt}$$

or
$$\frac{Id^2\theta}{dt^2} + \left(\frac{m}{R} + p\right)\frac{d\theta}{dt} + c\theta = 0$$

or
$$\frac{d^2\theta}{dt^2} + \frac{1}{I}\left(\frac{m}{R} + p\right)\frac{d\theta}{dt} + \frac{c\theta}{I} = 0$$

or
$$\frac{d^2\theta}{dt^2} + 2b\frac{d\theta}{dt} + K^2\theta = 0 \qquad ...(1)$$

where, $2b = \frac{1}{I}\left(\frac{m}{R} + p\right)$ ...(2)

and $K^2 = \frac{c}{I}$ ...(3)

Let the solution of this equation be $\theta = Ae^{\alpha t}$

Then, $\frac{d\theta}{dt} = A\alpha e^{\alpha t}$ and $\frac{d^2\theta}{dt^2} = A\alpha^2 e^{\alpha t}$.

Putting these values in equation (1),

$$\alpha^2 + 2b\alpha + K^2 = 0$$

or $$\alpha = -b \pm \sqrt{b^2 - K^2}$$

This gives,

$$\theta = A_1 e^{(-b + \sqrt{b^2 - K^2})t} + A_2 e^{(-b - \sqrt{b^2 - K^2})t}$$

where, A_2 and A_1 are constants.

The quantity $b^2 - K^2$ may be positive, zero or negative depending upon whether $b > K$, $b = K$ or $b < K$.

Case 1: When $b > K$, we have two real values of a, then solution of equation (1) is,

$$\theta = A_1 e^{\alpha 1t} + A_2 e^{\alpha 2t}$$

The α_1 and α_2 are negative. The deflection θ will go no decreasing with increasing time. *The motion is non-oscillating or dead beat and the oscillation is said to be over-damped.*

Case 2 : Critical damping : When $b = K$, the two values of a are equal to $-b$ and solution of (1) becomes

$$\theta = (A_1 + A_2)\, e^{-bt}$$
$$= A_3\, e^{-bt}$$

That the coil after deflection comes to rest in minimum possible time and the galvanometer is said to be *critically damped. However, the motion of the coil is just non-oscillatory*

Case 3 : When $b > K$, the quantity $\sqrt{b^2 - K^2}$ is imaginary so the values of a are complex *i.e.*,

$$\alpha = -b \pm jg. \text{ where } g^2 = K^2 - b^2$$

Hence, solution of equation (1) becomes,

$$\theta = A_1\, e^{(-b + jg)t} + A_2 e^{(-b - jg)t}$$

$$= e^{-bt}\,[A_1 e^{jgt} + A_2 e^{-jgt}]$$

$$= e^{-bt}\,[A_1\,\{\cos gt + j \sin gt\} + A_2\,\{\cos gt - j \sin gt\}]$$

$$= e^{-bt}\,[(A_1 + A_2) \cos gt + j\,(A_1 - A_2) \sin gt]$$

Putting $\quad A_1 + A_2 = B \sin \beta$

and $\quad j\,(A_1 - A_2) = B \cos \beta$

$\theta = Be^{-bt} \sin (gt + \beta) = Be^{-bt} \sin \{(\sqrt{K^2 - b^2})\, t + \beta\}$.

The motion is oscillatory or ballistic and period of oscillation,

$$T = \frac{2\pi}{g} = \frac{2\pi}{\sqrt{K^2 - b^2}} \qquad \text{...(4)}$$

In order that the whole charge may pass through the ballistic galvanometer before it has moved from its zero position, it is essential that its periodic time be large. Since we have,

$$T = 2\pi \sqrt{\frac{I}{c}}.$$

therefore, I should be large and c should be small. As we have

$$b < K$$

$$\frac{1}{2I}\left(\frac{m}{R} + p\right) < \frac{c}{I} < \frac{2\pi}{T}$$

Hence to make b small,

I should be large,

R should be large,

m should be small and so the coil should be wound round a nonconducting frame, *e.g.*, paper, bamboo or plastic. p should be small, and so the air resistance should be small and the fibre should be very fine.

We just see that the requirements of dead-beat galvanometer are just opposite to those for the ballistic b > K. Hence the opposite conditions should prevail. The coil is made of thick copper wire and has smaller number of turns so to have smaller resistance and is wound round a metallic or conducting frame.

SOLVED EXAMPLES

Example 1:

When 0.1 coul. of charge is passed through a moving coil B.G. a deflection of 30 mm is observed on a scale one meter away. Find the current sensitivity of the galvanometer if the time period of the coil is 10 seconds.

Solution:

From equation (1)

$$\text{Current sensitivity} = \frac{T}{2\pi} \times \text{charge sensitivity}$$

$$= \frac{10\text{sec}}{2\times 3.14} \times \frac{30\text{ mm}}{0.1\text{ Coul}}$$

$$= 477.5\text{ mm/amp.}$$

Example 2:

A condenser of capacity 0.2 μF is charged to 3 volts when discharged through a B.G., it gives a deflection of 10 cm. Calculate the current sensitivity of galvanometer, if its periodic time be 10 sec.

Solution:

$$\text{Charge sensitivity} = \frac{10}{0.2\times 3}\ \frac{\text{cm}}{\text{micro coulomb}}$$

$$= 16.6\ \frac{\text{cm}}{\text{micro coulomb}}$$

$$\text{Current sensitivity} = \frac{2\pi}{T}.\ \text{Charge sensitivity}$$

$$= \frac{2\times 3.14}{10\text{ sec}} \times \frac{16.6\text{ cm}}{\text{micro coul}}$$

$$= 10.4\text{ cm/micro amp.}$$

Example 3:

A circular coil of 100 turns has an effective radius 50 cm and carries a current 0.10 amp. Calculate the work done required to turn it in an external uniform magnetic field 1.5 weber/metre² through 180°.

Solution:

Let us take the coil be turned from position $\theta = 0$ to $\theta = 180°$, the work required is the difference in energy between the two positions, the potential energy $\vec{U} = -m.\vec{B}$

. Hence, work required $= U_e = 180\ {}^{-U}\theta = 0$

$$= -\text{mBCos } 180 - (-\text{mB Cos } 0)$$

$$= 2\text{mB}$$

But m = NiA

Hence, work required = 2 NiAB

$$= 2\text{Ni}(\pi r^2)\text{ B}$$

$$= 2 \times 100 \times (0.10) \times \pi \times (5 \times 10^{-2})1.5$$

$$= 0.24 \text{ joule}$$

Example 4:

A condenser charged to 2 volts is discharged through a ballistic galvanometer, when the corrected deflection is 1.6 cms and the current sensitivity is 4.54 × 10² mm/μA and the periodic time is 12 seconds. Calculate the capacity of the condenser.

Solution:

$$\text{Charge sensitivity} = \frac{T}{2\pi} \times \text{current sensitivity}$$

$$= \frac{12}{2 \times 3.14} \times 4.54 \times 10^2 \text{ mm/micro coul}$$

$$= \frac{12 \times 4.5 \times 10}{2 \times 3.14} \text{ cm/micro coul}$$

But $\quad \text{charge sensitivity} = \dfrac{\text{deflection}}{\text{charge}}$

or $\quad \text{Charge} = \dfrac{\text{deflection}}{\text{charge sensitivity}}$

$$= \frac{9.6 \text{ cm}}{\dfrac{12 \times 4.5 \times 10}{2 \times 3.14} \text{ cm/micro coul}}$$

or $\quad q = \dfrac{9.6 \times 2 \times 3.14}{12 \times 4.5 \times 10}\ \mu \text{ coul}$

$$= \frac{9.6\times2\times3.14\times10^{-6}}{12\times4.5\times10} \text{ coul}$$

But, $$q = c\,V$$

or $$c = \frac{q}{V} = \frac{9.6\times2\times3.14\times10^{-6}}{12\times4.5\times10\times2} \text{ Coul/volt}$$

$$= 20.16 \times 10^{-8} \text{ Farad}$$

$$= 0.2016\ \mu\text{F}.$$

Example 5:

The successive deflections to the right and lift of the mean position in the case of ballistic galvanometer are found to be 25.9 and 24.8 cms. Calculate the logarithmic decrement.

Solution:

As we know that $$\frac{\theta_1}{\theta_2} = \frac{\theta_2}{\theta_3} e^{\lambda}$$

$$e^{\lambda} = \frac{25}{24.9} = \frac{24.9}{24.8} = \frac{25+24.9}{24.9+24.8}$$

$$\lambda = 2.3 \log_{10} \frac{49.9}{49.7} = 0.0039$$

Example 6:

A condenser of 1,000 of is charged to a potential difference of 1 volt and then discharged through a B.G. The first throw an noted on a scale placed 1 meter away is 62.2 cms. If the time period of free vibrations be 10 seconds, and logarithmic decrement 0.02, calculate the ballistic constant of the galvanometer and its figure of merit.

Solution:

$$q = cV$$

$$= 1000 \times 10^{-12} \times 1 = 10^{-9} \text{ coul.}$$

and also $$q = K\theta_1\left(1+\frac{\lambda}{2}\right),$$

where, $$K = \frac{T}{2\pi}\frac{c}{NAB}$$

or, $$10^{-9} = K \times 62.2 \times 10^{-2} \times \left(1+\frac{0.02}{2}\right)$$

or $$K = \frac{10^{-9}}{62.2\times10^{-2}\times\left(\frac{2+0.02}{2}\right)}$$

$$= \frac{10^{-9}\times1000\times100}{622\times101}\ \frac{\text{Coul}}{\text{metre}}$$

Since, $$K = \frac{T}{2\pi}\left(\frac{c}{NAB}\right)$$

In fact, $\frac{c}{NAB}$ is the figure of merit

$$\frac{c}{NAB} = K.\ \frac{2\pi}{T} = \frac{10^{-9}\times1000\times100}{622\times101}\ \frac{2\pi}{10}$$

$$= 10^{-10} \text{ amp/radian,}$$

Let us consider a deflection of 1 mm on a scale, 100 cm away

Angular deflection $= \frac{10.3}{1} = \frac{1}{1000}$ radian

Current required to produce this deflection,

$$= 10^{-10} \text{ amp/radian} \times \frac{1}{1000} \text{ radian}$$

$$= 10^{-13} \text{ amp.}$$

Example 7:

A 50 ohm coil galvanometer can carry a current 10 MA. How can you convert it into

(a) voltmeter of range 500 volts

(b) ammeter of range 10 amp.

Solution:

(a) Using relation (1), *i.e.*,

$$V = i_g (R_m + R_g) \text{ or } R_m = \frac{V}{i_g} - R_g$$

$$i_g = 10MA = 10 \times 10–3 \text{ amp}$$

$$R_g = 50 \text{ ohms}$$

$$V = 500 \text{ volts}$$

hence, $$R_m = \frac{500}{10\times10^{-3}} - 50 = 49{,}950 \text{ ohms}$$

$$R_m = 49950, \text{ is connected in series.}$$

(b) Using relation (1), *i.e.*,

$$i = i_g\left(1+\frac{R_g}{R_s}\right)$$

given, $$i = 10 \text{ amp}$$

$$i_g = 10 \times 10^{-3} \text{ amp.}$$

$$R_g = 50 \text{ ohms}$$

Hence, $$10 = 10 \times 10^{-3}\left(1+\frac{50}{R_s}\right)$$

or $$\frac{10}{10\times10^{-3}} = \frac{R_s+50}{R_s}$$

or $$1+\frac{50}{R_s} = 1000$$

or $$\frac{50}{R_s} = 999$$

or $$R_s = \frac{50}{999} = 0.05005$$

R_s is shunt resistance.

Example 8:

A pointer type galvanometer of 20 ohm resistance has a linear scale of 50 divisions. It gives a deflection of 30 divisions when connected in series with a resistance of 2980 ohms and an accumulator of 2 volts. How will you convert it into an ammeter of range 3 ampere ?

Solution:

The current corresponding to 30 divisions by Ohm's law.

$$i_{30} = \frac{2}{2980+20} = \frac{2}{3000} \text{ amp.}$$

The maximum value of the current galvanometer can carry is corresponding to 50 divisions. Hence,

$$i_g = i_{50} = \frac{2}{3000} \times \frac{50}{30} \text{ amp.}$$

Now using the relation,

$$i = i_g \left(1 + \frac{R_g}{R_s}\right)$$

$$3 = \frac{2}{3000} \times \frac{50}{30}\left(1 + \frac{20}{R_s}\right)$$

or
$$\frac{3 \times 3000 \times 30}{2 \times 50} = 1 + \frac{20}{R_s}$$

or
$$\frac{20}{R_s} = \frac{3 \times 3000 \times 30}{2 \times 50} - 1$$

$$= 2700 - 1 = 2699$$

or
$$R_s = \frac{29}{2699}\ 0.00741 \text{ ohm (shunt)}$$

Example 9:

A copper strip 2.0 cm wide and 1.0 mm thick is placed in a magnetic field of magnitude 1.5 webers/metre^2 in the z-direction. If a current of 100 amp is set up in the strip along the x-direction, calculate the Hall potential difference across the strip given the number of electrons per cubic metre = 8.4 × 10^{28} metre^3.

Solution:

From equation (1),

$$E_y = \frac{j_x B_z}{Ne}$$

But $\quad E_y = \dfrac{V}{d}$ where V is the potential difference

and $\quad jx = \dfrac{i}{A} = \dfrac{i}{dh}$

where h is thickness.

Substituting these values, we get

$$\frac{V}{d} = \frac{i}{dh} \cdot \frac{B_z}{Ne}$$

$$V = \frac{iB_z}{Neh} = \frac{200 \text{ amp} \times 1.5 \text{ weber / metre}^2}{8.4\times10^{28/m^3} \times 1.6\times10^{-19} \text{ coul} \times 1.0\times10^3 \text{ m}}$$

$$= 2.2 \times 10^{-5} \text{ volt}$$

$$= 22 \text{ micro volt.}$$

Example 10:

A uniform magnetic field of magnitude 1.5 Weber/metre² points horizontally from south to north. A proton of energy 5.0 MeV moves vertically downward through this field. Calculate the force on it.

Solution:

Kinetic energy of the proton is

$$K = 5.0 \text{ MeV} = 5 \times 10^6 \text{ eV}$$

$$= 5 \times 10^6 \text{ eV} \times 1.6 \times 10^{-19} \text{ joule}$$

$$= 8.0 \times 10^{-13} \text{ joule.}$$

But $K = 1/2\ mv^2$, where m is the mass and v is the velocity, or

$$v = \sqrt{\frac{2K}{m}} = \sqrt{\frac{2\times8.0\times10^{-13}}{1.7\times10^{-27}}} = 3.1\times10^7 \text{ metres/sec.}$$

Now, according to equation (1.2),

$$F = qvB \sin\theta.$$

As given $\vec{B}$ is horizontal and proton moves vertically downward, hence angle between $\vec{v}$ and $\vec{B}$ is 90°,

therefore, $\sin\theta = \sin 90 = 1$

$F = qvB = (1.6 \times 10^{-19}\text{Coul}) \times (3.1 \times 10^7\text{metre/sec}) \times 1.5\text{weber/}$

metre^2

But, $\text{Weber} = \dfrac{\text{Newton.metre}}{\text{Ampere}}$

hence, $F = 7.4 \times 0^{-12}$ Netwons.

Since $\vec{F}$ is perpendicular to $\vec{v}$ and $\vec{B}$ from relation,

$$\vec{F} = q\vec{v} \times \vec{B},$$

the direction of this force is east.

Example 11:

A 10 cm long wire carrying a current of 10 amp. is held at an angle 30° with the direction of a uniform magnetic field of strength 1 weber/ metre². Calculate the force acting on the wire.

Solution:

According to equation (1.6), the force is

$$\vec{F} = i\vec{l} \times \vec{B}$$

$$\vec{F} = ilB \sin \theta.$$

Putting the given values,

$$F = 10 \times 10 \times 10^{-2} \times 1 \sin 30$$

$$= 0.5 \text{ Newton.}$$

EXERCISES

1. Obtain an expression for force on a magnetic dipole placed in a non-uniform magnetic field.
2. A one metre long wire carries a current of 10 amp. and makes an angle 30° with a uniform magnetic field B = 1.5 webre/m². What is the magnitude and direction of the force acting on the wire ?
3. In Bohr model of hydrogen atom the electron rotates around the nucleus in a circular path of radius 5.1×10^{-11} metre at a frequency v of 6.8×10^{15} revolutions/sec. The value of B set-up at centre of the orbit is 14 weber/m². Find the equivalent dipole moment.
4. Explain how magnetic induction $\vec{B}$ may be defined without using the concept of magnetic poles.
5. Obtain an expression for the force on the wire placed in magnetic field $\vec{B}$. Explain current element. Why $\vec{B}$ is also called as the magnetic flux density ?
6. What do you mean by flux of magnetic field? Calculate the magnetic flux over a closed surface. What is the physical significance of the result ?
7. An electron is moving with velocity $2\hat{i} + 3\hat{j}$ cm/sec. in an electric field of intensity $3\hat{i} + 6\hat{j} + 2\hat{k}$ and a magnetic field of

$2\hat{j} + 3\hat{k}$ Tasle. Find the magnitude and direction of Lorentz force acting on the electron.

8. The current sensitivity of a B.G. is 4.54×10^2 mm/micro amp. on a scale kept at a distance 1 metre. Calculate the charge sensitivity of galvanometer if the time period of the coil is 6.2 sec.

9. The current sensitivity of a B.G. is 2.2×10^{-9} amp. for a deflection of 1 mm on a scale kept at a distance of 1 metre. Calculate the charge sensitivity of the galvanometer if the time period of the coil = 6.2 sec.

2

Theory of Magnetism

INTRODUCTION

The fields of permanent magnets (outside the magnets) may be described either in terms of equivalent currents or equivalent poles*. Since the entire description of magnetism has been based on the assumption that are produced by moving charges, we are led to believe that the interpretation of the field of a permanent magnet in terms of electric charges giving rise to atomic currents is a more fundamental one than the concept of magnetic charges. The electrons have at least two motions in an atom giving rise to two kinds of currents listed below:

1. The currents due to orbital motion of the electrons in atoms.
2. The currents due to spinning of electrons about their own axes.

Such circulating currents were first postulated by Ampere, and are often called amperian currents.

The question now arises as the which is more important, the orbital motion of the electrons or the electronic spin. The electronic spin takes precedence in an explanation of the behaviour of magnetic materials.

In many atoms the electrons are completely paired, *i.e.,* for each electron spinning in one direction there is an electron spinning in the opposite direction. And the same situation exists in regard to the orbital motion of the electrons. Thus, the net current circulating about any given axis is zero and there are no amperian currents in such substances. Such materials exhibit very weak magnetic effect, they are called "non-magnetic". In other words, the electron spins and orbital motions are exactly balanced out, so that any particular atom has no average amperian current. Such substances when placed in magnetic field little extra amperian currents are generated inside the atom by induction, due to the changes in the angular velocity of the electrons in their orbits. According to Lenz's law, these currents are in such a direction as to oppose the applied field. This effect is known as "diamagnetism".

In a magnetic substance, however, the cancellation is incomplete and the amperian currents are not zero. Obviously, those substances whose atoms have an odd number of electrons must have at least one uncompensated spinning electron. These electrons have the same effect as tiny electric currents known as amperian currents. In other words, a magnetic material contains tiny amperian currents distributed throughout its entire volume. *Magnetisation consists in the alignment of these amperian currents.* If the magnetic field due to this amperian current has the same direction as the applied magnetic field, obviously the field is increased by the presence of the amperian current. This magnetic effect is known as "paramagnetism".

However, the paramagnetism arising from the spins of the electrons responsible for conduction in a metal constitutes an exception, we will not be discussing the phenomena here. Paramagnetism is generally fairly weak because lining up forces are relatively small compared with the forces from the thermal motions which try to change the order. In the case of iron there are as many as four uncompensated spinning electrons within each atom of molecule and consequently iron tends to be "strongly magnetic". Substances that behave like iron are called "ferromagnetic".

SOME MAGNETISATION POINTS

Non-ferromagnetic materials which produce only a small change in field, can be either diamagnetic or paramagnetic. If the field is decreased very slightly then the material is diamagnetic. If the field is increased slightly then the material is paramagnetic. The most easily observed characteristics of the different materials are their behaviours in strong non-uniform fields. If a specimen is placed in a magnet field, the field in the vicinity of the sample is changed. How much it is changed depends on the material and on the shape of the sample, but for most specimens it is not altered very much-typically only by about one part in 10^5. For certain materials, however, the field at some points nearby can be increased by a factor of one hundred or so over its value in the absence of the specimen. Such materials are called ferromagnetic. A ferromagnetic material is strongly pulled into the field. A diamagnetic material is repelled towards the field tree regions, but so weakly that sensitive apparatus is required to measure the repulsive force A paramagnetic material is pulled into the field, although the attraction is usually weak. This attraction or repulsion can be explained in terms of the interaction of magnetic dipole moments with the field in which the materials are placed. Certain atoms or molecules may have permanent magnetic dipole-moments.

If they do have permanent moments, the material is paramagnetic, other wise it is diamagnetic. If paramagnetic molecules (or materials) are placed in a magnetic field there is a net alignment of the dipoles in the field direction and the substance becomes magnetised. When a magnetic material is placed in a magnetic field, the atomic currents due to uncompensated spinning electrons are aligned and the process is called magnetisation. To be more specific, since circulating electron is equivalent to a circular coil which in turn is equivalent to a small magnet. Thus the alignment of atomic currents is same as alignment of small magnets because, it is sufficient to describe them by their magnetic moment which can be expressed either as $Q_m l$ or IA, (For detail see equivalence of circular coil to a magnet). *For simplicity we define the magnetisation as the alignment of elementary (or atomic) magnetic dipoles.*

In a non-uniform magnetic field a para-magnetic material will thus experience a force in the direction of increasing field. All atoms and molecules acquire an induced magnetic dipole moment when placed in a magnetic field. Induced magnetic dipole moments, unlike induced electric dipole-moments, are in a direction opposite to the field in the specimen. Thus, if the molecules of a substance have no permanent moments *i.e.,* the substance is diamagnetic, a sample inserted into a non-uniform magnetic field will experience a force in the direction of decreasing field strength. Clearly, the process of alignment of the dipoles as well as the phenomena by which atones or molecules acquire an induced magnetic deployments when a substance is placed in a magnetic field is known as magnetisation. Thus, any process by which a substance acquires (or develops) a net magnetic moment is magnetisation. If we want to know the origin of diamagnetism and paramagnetism we must know the origin of magnetic dipole-moment of an atom or molecule. There are two chief contributions to the magnetic dipole-moment of an atom or molecule.

Uniform Magnetisation

Magnetisation of any substance is uniform if all the atomic magnetic dipoles are uniformly distributed throughout the volume of the substance and are oriented in the same direction as the magnetising field. Thus, magnetisation is constant throughout the material. Magnetisation is uniform if the substance is *homogeneous and isotropic.*

Magnetisation Vector $\vec{M}$

If a specimen is placed in a magnetic field then by the process of magnetisation the atoms of molecules are magnetied on the atomic scale. Precisely, the magnetisation is the interaction of magnetic dipole moments with the field in which the materials are placed. This interaction (*i.e.,* the effect of magnetising field on the magnetic dipoles can conveniently be described by a quantity called magnetisation vector $\vec{M}$. The microscopic magnetisation can very from atom to atom but can be averaged over volume which contain very many atoms to give a function which is smoothly varying with position on a macroscopic scale. This function is called the magnetisation of the medium and is represented by symbol $\vec{M}$. It is defined as the magnetic dipole moment per unit volume, and is given by

$$\vec{M} = \frac{\text{Magnetic dipole moment}}{\text{volume}} \qquad ...(1)$$

Magnetisation at a point

The value of $\vec{M}$ in above equation (2) is an average for the volume. If magnetisation is uniform, the substance has a continuous distribution of infinitesimal atomic magnetic dipoles and the dipoles are of discrete, finite size. Then, for continuous magnetisation, the value of $\vec{M}$ at a point can be defined as the net dipole moment $\sum \vec{m}$ of a small volume Δv divided by the volume with the limit taken as Δv shrinks to zero around the point, thus,

$$\vec{M} = \lim_{\nabla v \to 0} \frac{\vec{\nabla}_m}{\Delta v} \qquad ...(2)$$

If the substance is not magnetised, the individual moments $\vec{m}$ will be oriented at random and the vector sum $\sum \vec{m}$ will be practically zero. In this case $\vec{M}$ is zero.

Units of $\vec{M}$.

Dimensionally eqn. (3) has the form,

$$\text{Magnetisation} = \frac{\text{Magnetic dipole moment}}{\text{volume}}$$

$$= \frac{\text{Current} \times \text{Area}}{\text{volume}} \qquad ...(3)$$

$$= \frac{\text{Amp. met}^2}{\text{met}^2} = \frac{\text{Amp.}}{\text{met.}}$$

Thus, the RMKS units of magnetisation is ampere per metre.

The currents equivalent to a magnetised body

In this section we will discuss how a magnetised body can be replaced by currents in order to determine the magnetic field arising from the magnetisation. Let us first estimate the equivalent current in a simple example *i.e.,* a bar magnet.

Consider a bar magnet in which the magnetisation is every where in the direction of the positive z-axis. Let us consider the volume Δv to be a small rectangular box as shown in Fig. 2.1. Since all the atomic currents within the box may be imagined to flow around the area $\Delta x\ \Delta y$, a resultant current (or fictitious current also amperian current) I_m may be considered to flow in one direction over a face. The magnetisation $\vec{M}$ is given at a point. By,

$$M = \frac{I_m\ \Delta x\ \Delta y}{\Delta x\ \Delta y\ \Delta z} = \frac{I_m}{\Delta z} = I' \qquad ...(4)$$

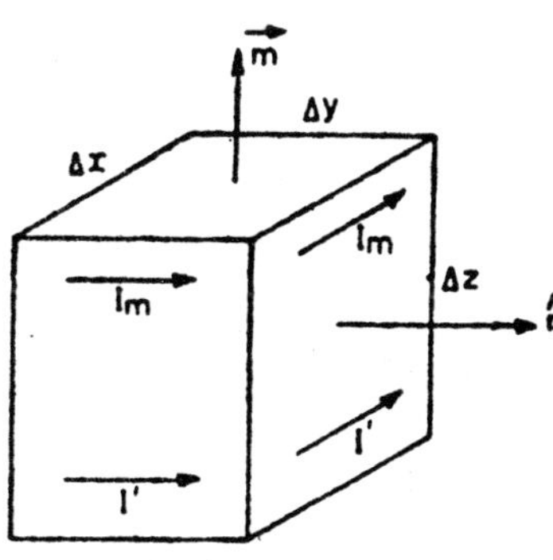

Fig. 2.1

Since magnetic dipole moment = current × area

where I' is the current flowing per unit length over the sides of the volume. Thus we must assume a surface current (I') at the surface of the magnetised volume numerically equal to the magnetisation. We can indicate the direction as well as the magnitude of this surface current by the vector equation,

$$\text{Surface current } I' = \frac{I_m}{\Delta z} = \vec{M} \times \hat{n} \qquad \text{...(5)}$$

where $\hat{n}$ is the out-drawn normal to the surface. We call I' also as the *linear current density.*

Non-uniform Magnetisation

In this case all the atomic magnetic dipoles are distributed in a non-uniform manner. *The magnetisation varies from point to point in the material.* $\vec{M}$ is a function of position in a non-uniformly magnetised material. Then the total magnetic moment,

$$\Sigma \vec{m} = \int_V \vec{m} \, dv \text{ amp. metre}^2 \qquad \text{...(6)}$$

where integration is carried out over the volume of the substance. Non-uniform magnetisation occurs when substances are *inhemogeneous and non-isotropic.* If at two points magnetisation is $\vec{M}$ and $\vec{M} + \Delta\vec{M}$. then

$$\text{Surface current} = \Delta\vec{M} \times \hat{n} \qquad \text{...(7)}$$

where $\hat{n}$ is normal to the surface.

If the magnetisation is inhomogenous *i.e.,* magnetisation varies from point to point in the material. There will not be the complete cancellation of the circulating currents between the boundaries of the magnetised material and there will be a net current flowing throughout the volume of the material.

Let us find the resultant current density. Consider two adjacent rectangular current loops 1 and 2 in a magnetised material. Since magnetisation is not uniform, it is different for the two loops. If $\vec{M}$ is magnetisation vector, *i.e.,* magnetic moment per unit volume. The moment ($\vec{m}$) of a single loop in Fig. 2.2 is given by,

$$\vec{m} = \vec{M} \, dx \, dy \, dz$$

where dx dy dz is the volume of the loop.

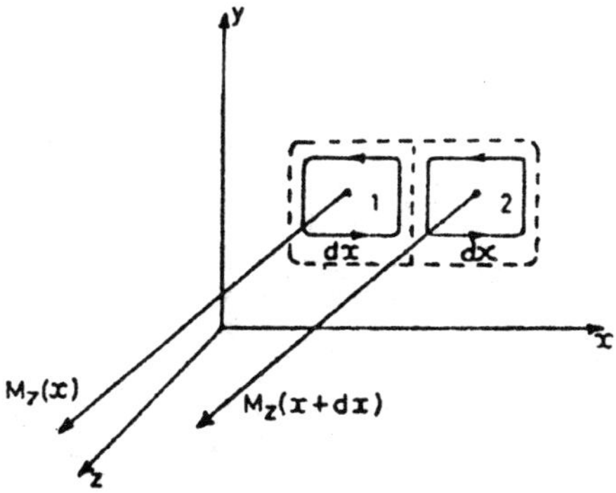

Fig. 2.2

Suppose rectangle has a magnetisation Mz in the z-direction. But we know that

Magnetic moment = current × area (IA)

Therefore, the current (amperian or magnetising) in the rectangle 1 for the considered case if denoted by I_{m1} is given by,

$$I_{m1} = \frac{M_z dx\,dy\,dz}{dx\,dy}$$

where dx dy is the area of the coil.

If magnetisation is changing at the rate of $\frac{\delta M_z}{\delta x}$ along x-axis then magnetisation at a point where second rectangle is situated is

$$= M_z + \left(\frac{\delta M_z}{\delta x}\right)dx$$

Therefore the current in the rectangle two is

$$I_{m2} = \left[M_z + \left(\frac{\delta M_z}{\delta x}\right)dx\right]\frac{dx\,dy\,dz}{dx\,dy}$$

The difference $I_{m1} - I_{m2}$ is the net current at the mutual boundary of rectangles 1 and 2. If it is denoted by Im then,

$$I_m = I_{m1} - I_{m2} = -\frac{\delta M_z}{\delta x}dx\,dz$$

the direction is determined by equation (8) by vector product, rule Im is normal to the plane of $\vec{M}$ and $\hat{n}$ *i.e.,* in the y-direction.

However, if $\vec{M}_x$ *i.e.,* x component of $\vec{M}$ varied with z, then we should also have a component of Im in y-direction, *i.e.,*

$$= \frac{\delta M_x}{\delta z} dx\, dx \qquad ...(8)$$

Thus combining these two we have the resultant current in y-direction,

$$I_{my} = \left[\frac{\delta M_x}{\delta x} - \frac{\delta M_z}{\delta x}\right] dx\, dz$$

dx dz is the area of coil in x-z plane. Therefore the current density in y direction,

$$J_{my} = \frac{I_{my}}{dx\, dz} = \left[\frac{\delta M_x}{\delta z} - \frac{\delta M_z}{\delta x}\right]$$

$$= (\text{curl } M)y$$

We have similar expressions for x and z components *i.e.,*

$$J_{mx} = (\text{curl } M)_x$$

and $$J_{mz} = (\text{curl } M)_z$$

Hence, we can generalise,

$$\vec{J}_m = \text{curl } \vec{M} \qquad ...(9)$$

Important Points

1. In fact, the amperian currents (or atomic currents) are fictitious and are replaced by the effective magnetising currents that give the same magnetic effect of the material, we shall discuss this in the next article.
2. The magnetisation $\vec{M}$ is analogous to the electric polarisation $\vec{P}$. *The amperian current in magnetic materials corresponds to the induced charges in the dielectrics.*

Three Magnetic Vectors

If we place a magnetic material in a magnetic field the elementary magnetic dipoles will be aligned and set up their own field, which will modify the original field. Let us consider a toroidal coil (Rowland ring) carrying a current I_o in it. The ring is so constructed that its core can be easily replaced. If the current in the winding is constant the value of B will be greater when the core is in place than without it.

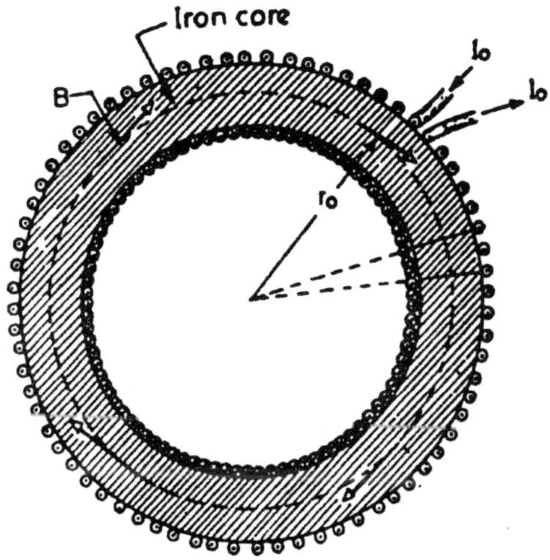

(a)

Fig. 2.3

The large value of B is due to magnetisation of the material of the core. If I_0 is the current through toroid of radius r_0, N_0 are the number of turns then, according to Ampere's law, the value of $\vec{B}$ without iron core:

$$\oint \vec{B}.d\vec{l} = \mu_0 I \quad ...(10)$$

$$(B)\ 2\pi r_0 = \mu_0 N_0 I_0 \quad ...(11)$$

Let us now put the core. The value of $\vec{B}$ determined experimentally is different when an iron core is placed than without it. As I_0 is the same then according to Ampere's law B must be same, which is not in agreement with experiment. *Thus Ampere's law fails.*

However, the value of $\vec{B}$ can be increased and equal to its value with core, even i the absence of the core by increasing the current in the toroid say by an amount I_{mo}. *The magnetisation of the iron core is thus equivalent in its effect on B to such a hypothetical current increase.* We modify the Ampere's law by adding magnetic current term I_m for fictitious amperian (or atomic) currents. To make it clear we can say as the Gauss's law was modified in the presence of dielectric, similarly the Ampere's law is modified in the presence or magnetic material, *i.e.,*

$$\oint \vec{B}.d\vec{l} = \mu_o (I + I_m) \quad ...(12)$$

or $$(B)\, 2\pi r_o = \mu_o N_o I_o + \mu_o (N_o\, I_{mo}) \quad ...(13)$$

In fact I_{mo} is related to the magnetisation $\vec{M}$, see eq. (10). *It is the value of the current which has the magnetic dipole moment equal to the magnetic moment produced due to the alignment of atomic dipoles.* If A is the area of the toroid, then the magnetic moment for the element dl of the toroid,

$$d\vec{m} = \vec{M}\,(Adl)$$

But magnetic moment is defined by,

$$m = NIA$$

where N are the number of turns, I is the current through the loop. For element dl, the number of turns

= Number of turns per unit length Xdl

$$= \frac{N_o}{2\pi r_o} dl$$

therefore the magnitude of magnetic moment, for element dl

$$M(Adl) = \frac{N_o}{2\pi r_o} dl.\, I_{mo} A$$

or $$N_o I_{mo} = M\, 2\pi r_o \quad ...(14)$$

Substituting this value of No Imo in equation (14), we get,

$$(B)\, 2\pi r_o = \mu_o (N_o I_o) + \mu_o (M\, 2\pi r_o)$$

or in general $\oint \vec{B}.d\vec{l} = \mu_o I + \mu_o \oint \vec{B}.d\vec{l}$

or, $$\oint \left(\frac{\vec{B} - \mu_o \vec{M}}{\mu_o} \right).d\vec{l} = I \quad ...(15)$$

The quantity $\frac{\vec{B} - \mu_0 \vec{M}}{\mu_0}$ is given a special name *magnetic field strength* $\vec{H}$, hence,

$$\vec{H} = \frac{\vec{B} - \mu_o \vec{M}}{\mu_o} \quad ...(16)$$

or, $$\vec{B} = \mu_o \vec{H} + \mu_o \vec{M} \quad ...(17)$$

or $$\vec{B} = \mu_o \left(1 + \frac{\vec{M}}{\vec{H}}\right)\vec{H} \quad ...(18)$$

Restatement of the Ampere's law:

Ampere's law, in the presence of magnetic material is given by eqns. (17) and (18), *i.e.,*

$$\oint \vec{H}.d\vec{l} = I \quad ...(19)$$

where I is true current and does not include the magnetising current. The equation states the most important property of field intensity $\vec{H}$ – *the line integral of $\vec{H}$ around any closed path equals the total conduction current encircled regardless of the presence of magnetised material.*

To find $\vec{H}$ in our toroid case, let us apply Ampere's law as given in equation (20).

$$H\,(2\pi r_0) = N_0 I_1$$

or, $$H = \left(\frac{N_0}{2\pi r_0}\right) I_o = n.I_o$$

where n are the number of turns per unit length.

Magnetising force or intensity $\vec{H}$

The relation $\oint \vec{B}.d\vec{l} = \mu I$ shows that the value of $\vec{B}$ depends on the permeability of the medium. From relation $\oint \vec{H}.d\vec{l} = I$, it can be seen that $\vec{H}$ does not depend on permeability m but only the currents. $\vec{H}$ is independent, therefore, of the medium in which the conductors are situated. It is this reason that $\vec{H}$ is regarded as being due to the currents only. $\vec{H}$ is then a "cause" which gives rise to a flux density $\vec{B}$ given by $\mu\vec{H}$, and so $\vec{B}$ is dependent on the medium used.

Because of this interpretation, $\vec{H}$ is often called the "magnetising force" or magnetising intensity. $\vec{H}$ is a vector in the direction of $\vec{B}$. The term magnetic intensity is not particularly appropriate since it implies that $\vec{H}$ is analogous to the electric field intensity $\vec{E}$, which is not precisely correct since in electric field $\vec{E}$ is defined in terms of force, whereas in magnetic field it is $\vec{B}$ that is defined in terms of force. Therefore, the name magnetising force is sometimes used for $\vec{H}$ which is more appropriate.

Unit of $\vec{H}$

The unit of $\vec{H}$ is obtained by equation (2.24) and is Ampere-turn per metre.

CONCLUSIONS ABOUT $\vec{B}.\vec{H}$ AND $\vec{M}$

Here, we are summarising the previous results.

(1) First property of $\vec{B}$ is derived from Gauss's law, *i.e.,* div $\vec{B}$ = 0, which means $\vec{B}$ lines are closed circles and do not have sources or sinks. In other words absence of monopole.

(2) Second property of $\vec{B}$ is expressed by the Ampere's law $\oint \vec{B}.d\vec{l}$ = $m_0 I$, which means $\vec{B}$ is *non-conservative* field *i.e.,* do not have sources or sinks in contrast to the electric field $\vec{E}$ which originates from charges..

(3) We have restated Ampere's law as,

$$\oint \vec{H}.d\vec{l} = I \qquad ...(1)$$

where I is the true conduction current. According to this relation $\vec{H}$ *is also a non-conservative field.*

By Stoke's theorem,

$$\oint \vec{H}.d\vec{l} = \int_S \text{Curl } \vec{H}.d\vec{S} \qquad ...(2)$$

And current,

$$I = \int_S \vec{J}.d\vec{S} \text{ (see eqn. 8.17)} \qquad ...(3)$$

By comparison of equations A, B and C we get,

$$\text{Curl } \vec{H} = \vec{J} \qquad ...(4)$$

This is the basic equation valid everywhere.

(4) We have established the relation

$$\oint \vec{M}.d\vec{l} = I_m$$

where I_m is amperian current, equation states that the line integral of the magnetisation around a closed path is equal to the total amperian currents through the surface encircled by the path.

Now using Stoke's theorem,

$$\oint \vec{M}.d\vec{l} = \int_S \nabla \times \vec{M}.d\vec{S}$$

and
$$I_m = \int_S \vec{J}_m.d\vec{S}$$

therefore, $\nabla \times \vec{M} = \vec{J}_m$ ...(5)

which is the same as proved previously in equation (5)

(5) **Important point** to be noted is that sometimes the relation, $\vec{B}$ = mo $(\vec{H} + \vec{M})$ is taken as,

$$\vec{B} = \vec{H} + \mu_0 \vec{M}$$

Then in RMKS, $\vec{H}$ will have the same unit as $\vec{B}$ *i.e.,* weber/metre^2. However in our relation $\vec{B} = \mu_o (\vec{H} + \vec{M})$, $\vec{H}$ is expressed in the same unit as $\vec{M}$ *i.e.,* ampere per metre. *This unit is more convenient in engineering for designing transformers, magnets etc.*

Further, in our relation $\vec{B}$ and $\vec{H}$ are expressed in different units. There is no difficulty with this definition of $\vec{H}$ in RMKS units. Hence we prefer relation $\vec{B} = \mu_o (\vec{H} + \vec{M})$ then to $\vec{B} = \vec{H} + \mu_0 \vec{M}$.

PERMEABILITY AND SUSCEPTIBILITY RELATIVE PERMEABILITY

We can measure $\vec{B}$, $\vec{H}$ and $\vec{M}$ for different magnetic materials For non-ferromagnetic that is, for paramagnetic and diamagnet materials it is found experimentally that $\vec{B}$ is directly proportional to $\vec{H}$

or $\vec{B} \propto \vec{H}$

or $\vec{B} = \mu_r \mu_o \vec{H}$ ...(1)

where μ_r (sometimes written as K_m) is the relative permeability of the magnetic medium. $\vec{B}$ is vector in the direction of $\vec{H}$. Putting the value of $\vec{B}$ from eqn. (1) in eqn. (2), we get,

$$\mu_r \mu o \vec{H} = \mu_o \vec{H} + \mu_0 \vec{M}$$

or $\vec{M} = (\mu_r - 1) \vec{H}$ ...(2)

Now in vacuum, the magnetisation $\vec{M}$ is zero, because there are no magnetic dipoles present, hence putting $\vec{M} = 0$, equation (2) reduces,

$$\vec{B} = \mu_o \vec{H} \qquad ...(3)$$

Since in vacuum $\vec{M} = 0$, hence from equation (4)

$$(\mu_r - 1) = 0$$

or $\quad \mu_r = 1$ for vacuum

Magnetic susceptibility

For para and diamagnetic substances the magnetisation is proportional to the magnetising field H,

$$M \propto H$$

or
$$\vec{M} = \chi_m \vec{H} \qquad ...(4)$$

where χ_m is known as the magnetic susceptibility.

Since, $\chi_m = \frac{M}{H}$. It may be defined as ratio of magnetisation to the magnetising field. Comparing equation (3) and (4), we get

$$\chi_m = \mu_r - 1$$

or
$$\mu_r = 1 + \chi_m \qquad ...(5)$$

The magnetic susceptibility of the medium is the measure of the capability of the medium to take up magnetisation. It is analogous to dielectric susceptibility.

Permeability

Substituting equation (5) in eqn. (4), we get,

$$\vec{B} = (1 + \chi_m) \mu_o \vec{H}$$

or
$$\vec{B} = \mu \vec{H} \qquad ...(6)$$

where $\mu = \mu_o (1 + \chi_m)$, is the permeability of the medium, the dimensionless quantity,

$$\frac{\mu}{\mu_o} = 1 + \chi_m = \mu_r \text{ (by eqn 5)} \qquad ...(7)$$

or
$$\mu = \mu_o \left(1 + \frac{M}{H}\right), \sin \chi_m = \frac{M}{H} \qquad ...(8)$$

where μ_r is relative permeability of the medium and is defined as the ratio of the number of field lines per square metre taken at right angles in the medium to that in air or free space.

The permeability μ *measures the degree of penetration of magnetic field through the substance or in other words it measures the capacity of the substance to take magnetisation.* In isotropic medium $\vec{M}$ and $\vec{H}$ are in the same direction, hence m is scalar. However, in non-isotropic media, $\vec{M}$ and $\vec{H}$ are, in general not in the same direction and m is not a scalar but becomes a nine components quantity or tensor. Therefore relation (9) is general relation and (10) is concise expression only for isotropic media.

Units

From equation (9), the units of μ

$$= \frac{\text{Units of B}}{\text{Units of H}}$$

$$= \frac{\text{Webers' metre}^2}{\text{Ampere / metre}}$$

$$= \frac{\text{Webers}}{\text{Ampere. metre}},$$

since henry $= \dfrac{\text{Webers}}{\text{Ampere'}}$

Magnetic Properties of Matter

$$= \frac{\text{Henrys}}{\text{Metre}}$$

or from equation (2.33), the unit of μ is same as mo

$$= \frac{\text{Weber}}{\text{Ampere} \cdot \text{metre}} = \frac{\text{Henrys}}{\text{metre}}$$

Since μ_r is ratio it is dimensionless quantity. From eqn. (2) since $\vec{M}$ and $\vec{H}$ are expressed in the same unit, *i.e.*, ampere/metre, the ratio, χ_m is dimensionless quantity.

REPRESENTATION OF $\vec{B}$, $\vec{H}$ AND $\vec{M}$

Since magnetisation occurs inside the materials. According to equation (1),

$$\vec{B} = \mu_o \vec{H} + \mu_o \vec{M}$$

At the boundary $\vec{H}$ is reversed. Thus within the magnets $\vec{M}$ and $\vec{H}$ point in opposite direction, as shown in Fig. 2.4. How $\vec{H}$ is reversed? Let us examine it as follows.

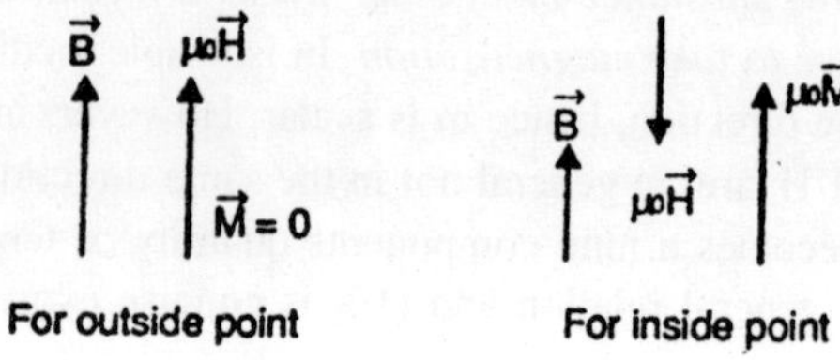

Fig. 2.4

Magnetic Scalar Potential

Taking the divergence of equation (1), *i.e.*,

$$\vec{B} = \mu_o \vec{H} + \mu_o \vec{M} \quad ...(1)$$

$$\nabla.\vec{B} = \nabla.\mu_o \vec{H} + \nabla.\mu_o \vec{M}$$

Since $\nabla, \vec{B} = 0$, we have,

$$\nabla.\vec{H} = -\nabla.\vec{M} \quad ...(2)$$

Since $\vec{M}$ and $\vec{H}$ having the dimensions of current/length, (ampere/metre). Then $\nabla.\vec{H}$ or $\nabla.\vec{M}$ have the dimensions as current per area (ampere/metre^2) which can be conveniently expressed as,

$$\frac{\text{Ampere. metres}}{\text{Metres}^2} = \frac{\text{Pole strength}}{\text{Volume}}$$

As we have proved that a current carrying coil is equivalent to a magnet (*i.e.*, IA = Q_{ml} where Q_m is the pole strength), hence dimensions of pole strength Q_m are (current × length), *i.e.*, (Amp. metre).

Therefore $\nabla \vec{H}$. or $\nabla.\vec{M}$ can also be expressed as pole strength per volume.

We can write,

$$\nabla.\vec{H} = -\nabla.\vec{M} = \rho_m \quad ...(3)$$

where ρ_m = pole volume density $\left(\frac{\text{Ampere. metres}}{\text{metres}^2}\right)$ the quantity ρ_m is like the polarisation volume density ρ_p.

If the divergence of a vector field is not zero, means the field has a source, or place of origin. Let us recall the polarised dielectric case, *i.e.*, $\nabla.\vec{p} = \rho_p$, which indicates that the polarisation field originates on the polarisation charge at the dielectric surface. In an analogous manner equation (2) *indicates that the* $\vec{H}$ *field originates where the magnetisation field* $\vec{M}$ *ends and the* $\vec{H}$ *field ends where the* $\vec{M}$ *field originates.* This occurs at the surface of the substance. Thus, if we consider points entirely outside of the current regions, and choose a path which do not enclose any current, thus when no current is enclosed,

$$\oint \vec{H}.d\vec{l} - 0 \qquad \text{(by Ampere's law)}$$

Under this condition $\vec{H}$ *is conservative field* and can then be derived from a scalar magnetic potential function* (V_m).

Thus when path does not enclose current, the above relation by Stoke's theorem can be written as

$$\nabla \times \vec{H} - 0$$

Then $\vec{H}$ can be expressed as negative gradient of a scalar in the same manner as in electrostatics $\nabla \times \vec{E} = 0$, then $\vec{E} = -\nabla V$.

$$\vec{H} = -\nabla V_m \qquad ...(4)$$

Taking div. of (4), we get,

$$\nabla.\vec{H} = -\nabla^2 V_m$$

Therefore, from equation (5),

$$\nabla^2 V_m = \nabla.\vec{M} \qquad ...(5)$$

This indicates that magnetic scalar potential V_m is related to source of magnetisation. *This is the "Poisson's equation" solution of which gives the magnetic scalar potential in current free regions.*

Vector Potential

Taking the curl of the relation,

$$\vec{B} = \mu_o \vec{H} + \mu_o \vec{M}$$

$$\nabla \times \vec{B} = \mu_o \nabla \times \vec{H} + \mu_o \nabla \times \vec{M} \qquad ...(6)$$

when there is no magnetisation, equation (6) reduces to,

$$\nabla \times \vec{B} = \mu_o \nabla \times \vec{H}, \text{ [since } \nabla \times \vec{H} = \vec{J} \text{ by eqn.} \qquad ...(7)]$$

$$= \mu_o \vec{J} \quad ...(8)$$

From relation (6), *the curl of* $\vec{M}$ *has the dimensions of current density (amp./metre*^{2}*) and represents the equivalent current of density* $\vec{J}_m$ (amp./metre^2) *flowing, in a very thin layer around the surface of a uniformly magnetised substance (as rod).*

Hence $\nabla \times \vec{M} = \vec{J}_m$ again the same conclusion as given in equation (7) and (8).

Thus equation (7) reduces,

$$\nabla \times \vec{B} = \mu_o (\vec{J} + \vec{J}_m) \quad ...(9)$$

where $\vec{J}$ = actual current density, as in a current carrying wire (amp./metre^2).

$\vec{J}_m$ = equivalent current density, as at the surface of a magnetised substance (bar) amp./metre^2

Potential due to a Magnetisation

We have defined the vector potential as,

$$\vec{B} = \nabla \times \vec{A}$$

and, $\vec{A} = \frac{\mu_o}{4\pi} \int \frac{\vec{J}}{r} dv'$ see eqn. (10)

If both conduction and magnetisation currents are present, we have

$$A = \frac{\mu_o}{4\pi} \int_o \frac{\vec{J} + \vec{J}_m}{r} dv' \quad ...(11)$$

where, $\vec{J} = \nabla \times \vec{H}$ amp./metre^2

$\vec{J}_m = \nabla \times \vec{M}$ amp./metre^2

COMMENT ON THE FIELD $\vec{H}$

We have restated Ampere's law as,

$$\oint \vec{H} . d\vec{l} = I$$

From the discussion of the preceding section it is clear that $\vec{H}$ can be regarded as sum of two components.

(1) H_1 due to true currents and it is non-conservative and

$$\oint \vec{H}_I . d\vec{l} = I$$

By non-conservative means it does not have sources of sinks.

(2) However, if we consider the magnetisation and if H_c is the field due to magnetisation, then it is a conservative field that satisfies the relation,

$$\oint \vec{H}_c . d\vec{l} = 0$$

In fact H_c represents the conservative field due to fictitious point charges those are assumed when substance is magnetised and then we have the magnetic scalar potential only in current free regions (As when $\oint \vec{H}.d\vec{l} = 0$, we express $\vec{E} = -\nabla V$).

BOUNDARY CONDITIONS AT THE SURFACE OF A MAGNETIC MATERIAL

Let us determine the relation between the values of $\vec{B}$ and $\vec{H}$ inside a magnetic material to the values of $\vec{B}$ and $\vec{H}$ just outside the material. The relation is obtained by two basic laws, the first is that the field $\vec{B}$ consists of closed lines, *i.e.*, the flux of $\vec{B}$ through a closed surface is zero which is precisely Gauss's law,

$$\nabla . \vec{B} = 0$$

or $$\oint \vec{B}.d\vec{l} = 0 \quad ...(1)$$

The second basic law is Ampere's law, *i.e.*,

$$\oint \vec{H}.d\vec{l} = I \quad ...(2)$$

Applying equation (1) to the cylindrical surface of height h shown in Fig. 2.5 we obtain,

$$B_{n1} \nabla S - B_{n2} \nabla S = 0 \quad ...(3)$$

where Bn1 is the average component of $\vec{B}$ normal to top face of the cylinder and $B_n{}^2$ is the average component of $\vec{B}$ normal to the lower face in inward direction, that is $B_n{}^2$ and ∇S are oppositely directed.

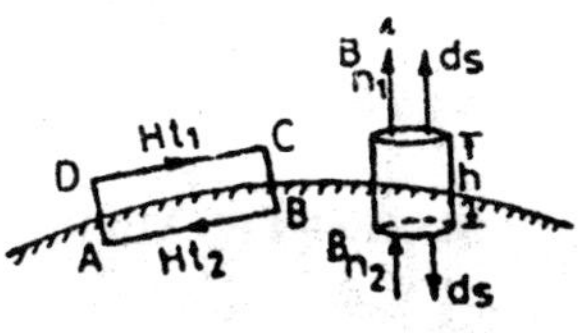

Fig. 2.5

The integral of $\vec{B}$ over the curved surface of the cylinder is neglected because by making height h approaching zero, integral zero even though there may be finite of $\vec{B}$ normal to the surface, by eq. (4)

$$B_n^1 = B_n^2 \quad ...(4)$$

Therefore, the normal component of $\vec{B}$ inside the substance is the same as the normal component of $\vec{B}$ just outside the substance. If the boundary separates two media of different permeability, even then equation (4) is valid. *Thus, the normal component of the* $\vec{B}$ *is continuous across the boundary between the two media.* Let us now apply Ampere's law to the path ABCD, AB = $-\nabla_x$. Let the average value of $\vec{H}$ tangent to the boundary for upper half is H_{t1} and for lower half path H_{t2}, then according to Ampere's law,

$$H_{t1}\,\nabla_x - H_{t2}\,\nabla_x = I$$

where I is current enclosed by ABCD. The integral of $\vec{H}$ over the segments BC and DA becomes zero even though a finite field may exist normal to the boundary by choosing BC and DA as small as possible and approach thing zero. The second term is negative because H_{t2} and ∇_x are oppositely directed.

$$H_{t1} - H_{t2} = \frac{I}{\nabla x} = I' \text{ amp/metre} \quad ...(5)$$

where I' is the linear density of current *i.e.,* current per unit length flowing in an infinitesimally thin sheet at the surface. According to equation (5), the change in tangential component of $\vec{H}$ across a boundary is equal in magnitude to the linear current density. If I' = 0

$$H_{t1} = H_{t2} \quad ...(6)$$

According to equation (6) the tangential component of $\vec{H}$ inside the magnetic material is the same as the tangential component of $\vec{H}$ just outside the material. If the boundary is separating two media of different permeability, even then equation (6) is valid. Thus, the tangential components of $\vec{H}$ are continuous across the boundary between two media provided the boundary has no current sheet of infinitesimal thickness.

CLASSIFICATION OF MAGNETIC MATERIALS

Faraday divided the magnetic materials into three groups in terms of susceptibility χ_m and relative permeability μ_r.

Paramagnetic Substances

For paramagnetic substances $\vec{M} = \chi_m \vec{H}$ and susceptibility cm is positive and less than unity and the relation $\mu_r = 1 + \chi_m$ indicates that relative permeability is slightly greater than unity. Susceptibility χ_m is independent of the field, when $\vec{H} = 0$ $\vec{M} = 0$ in this case. Paramagnetic substance when placed in magnetic field *acquire a feeble induced magnetism in the same direction as the external field as* $\chi_r > 1$. In a non-uniform field it will experience an attractive force toward the strongest point. The susceptibility cm varies inversely with the absolute temperature, mathematically,

$$\chi_m \propto \frac{1}{T}$$

or $$\chi_m T = \text{Const.} \quad ...(1)$$

Examples:

Some examples of paramagnetic substances are oxygen, platinum, aluminium, chromium, manganese and solutions of iron and nickel.

Diamagnetic Substances

Such substances acquire feeble induced magnetism in a direction opposite to the external field. In non-uniform field, they have tendency to move from stronger to weaker parts of the field. The magnetisation M is proportional to the magnetising field,

$$\vec{M} = \chi_m \vec{H}$$

Here, the susceptibility χ_m is independent of the temperature. χ_m is negative and relative permeability μ_r is slightly less than unity *i.e.*, $\mu_r < 1$. *Examples* of diamagnetic substances are bismuth, antimony, copper, mercury, gold, water, air, hydrogen and alcohol.

Ferromagnetic Substances

For ferromagnetic substances, the susceptibility χ_m is positive and is larger than para magnetic substances and the relative permeability is also much greater than unity. However, the relation $\vec{M} = \mu_c \vec{H}$ (*i.e.*, the magnetisation proportional to the field) does not hold for ferromagnetic materials. The ferromagnetic substances having spontaneous magnetisation even without the field, when $\vec{H} = 0$, $M \neq 0$ in this case. Due to greater value of μ_r, they are able to acquire high degree of magnetisation even they can develop north and south pole.

Distinction between Para and Ferro

Susceptibility decreases with increase in temperature. If the temperature is increased above a certain critical value, called Curie temperature the substance loses its ferromagnetism, that is, the spontaneous magnetisation disappears and substance becomes paramagnetic. Above the Curie temperature the susceptibility follows the Curie Weiss law, *i.e.*,

$$m = \frac{C}{T - \theta} \qquad ...(2)$$

where C is Curie-constant and temperature θ is Curie-temperature. *Hence we conclude that all ferromagnetism are basically paramagnetic but all para are not ferromagnetics. Above the curie temperature substance is paramagnetic and below it, it is ferromagnetic.*

Examples:

The five elements iron, cobalt, nickel, Gadolinium and dysprosium and variety of alloys of these and other elements.

MAGNETISATION CURVES (B – H CURVES)

The permeability μ of a substance, *i.e.*,

$$\mu = \frac{B}{H} = \mu_r \mu_o$$

But the permeability μ of the ratio B/H is not a constant for ferromagnetic materials in other words B is not proportional to $\vec{H}$. Therefore, in order to know the relation of $\vec{B}$ to $\vec{H}$ a graph showing the variation of B with H is plotted by taking B as ordinate and H as abscissa. The line or curve showing B as a function of H as such B-H curve is called magnetisation curve.

Ballistic Method for Plotting H-H Curve

To plot a magnetisation curve for an iron sample, take the sample in the form of a ring. A primary coil P is wound uniformly and closely spaced over the ring, forming an iron-cored toroid as in Fig. 2.6. The ends of the primary coil are connected to a battery B_1, through a resistance box R_1 and a reversing key K_1.

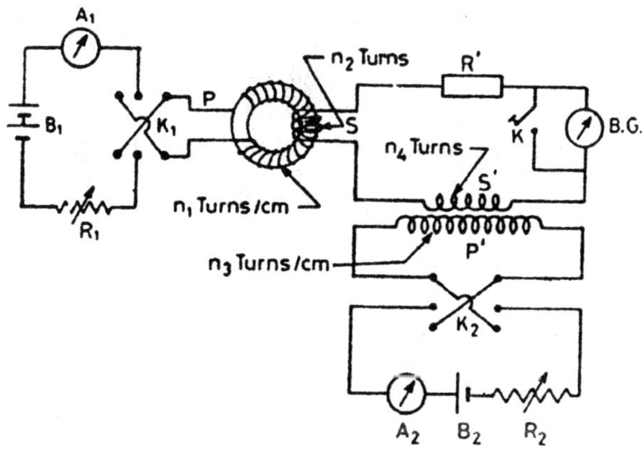

Fig. 2.6

The small coil S called the secondary coil, is wound over the primary coil P and is connected to ballistic galvanometer through a rheostat R' and secondary coil S'. The secondary coil S' itself wound over a primary P', which is further connected to battery B_2 through a reversing key K_2 (a commutator) for calibrating the B.G.

First the reversing key K_1 is switched on. A Transformers and A.C. Bridges I passes through the primary coil P. If the number of turns per unit length are n_1. Then, the value of H applied to the ring is,

$$H = n_2 I \text{ amp/metre or amp. turns/metre} \qquad ...(1)$$

(Since, for toroid $B = \mu_0 n_1 I$ and there is no magnetisation so far hence $\vec{M} = 0$, then relation $\vec{B} = \mu_0 \vec{H} + \mu_0 \vec{M}$ is $B = \mu_0 H$, hence $\vec{H} = \dfrac{B}{\mu_0} = n_1 I$).

This value of H applied to the ring may be called the "magnetising force". The magnetic induction B in the ring may be regarded as the result of applied field H and is measured by placing the secondary coil S over the ring. The magnetisation of the specimen develops a magnetic flux density (or magnetic induction) B inside the ring which is transferred to the secondary S. This transfer of flux to the secondary coil causes an induced emf. Suppose the flux density developed in the ring in the time dt = dB

Therefore, the flux developed pet turn = AdB where A is the area of cross-section of the ring. The flux linked with secondary = $dBAn_2$ where n_2 are the number of turns of the secondary.

Since initially flux in the secondary is zero.

The change in magnetic flux in the secondary = $d\phi = dB.An_2$

Rate of change of flux $\frac{d\phi}{dt} = An_2 \frac{dB}{dt}$

According to Faraday's law $\frac{d\phi}{dt}$ is equal to the induced emf in the secondary.

$$\xi = An_2 \frac{dB}{dt} \text{ volts}$$

And corresponding current in the secondary,

$$I_2 = \frac{\xi}{R} = \frac{An_2}{R}\frac{dB}{dt} \text{ amps}$$

where R is the total resistance of the galvanometer circuit.

Since current is rate of flow of change, hence,

$$\frac{dq}{dt} = \frac{An_2}{R}\frac{dB}{dt}$$

$$dq = \frac{An_2}{R} dB \text{ Couls}$$

Since initially magnetic induction B was zero it rises as current increases from zero to a saturation value B. Thus total charge developed in the secondary,

$$q = \int_O^B dq = \int_O^B \frac{An_2}{r} dB = \frac{An_2 R}{R} \text{ Coul}$$

Now if θ is the first throw in B.G. the,

$$q = K\theta\left(1+\frac{\lambda}{2}\right), \qquad \text{K is B.G. constant,}$$

from these two expressions, we get'

$$\frac{An_2 B}{R} = K\theta\left(1+\frac{\lambda}{2}\right) \qquad ...(2)$$

Determination of Galvanometer Constant

The value of K can be determined by the use of a standard solenoid. A known current I' is passed through the primary of the standard solenoid and suddenly this will cause change in magnetic flux is calculated as follows. The magnetic field in the primary P' = n_2I' where n_3 are the number of turns per unit length of the primary P'.

This creates a magnetic flux through the secondary,

$$\phi' = \mu_o n_3 I' \propto n_4$$

where $\propto$ is the area of cross-section of the secondary S' and n_4 are the number of turns in the secondary S'. This change in flux will give rise to an induced emf in the secondary hence proceeding in the manner as above, we get,

$$\frac{\mu_o n_3 I' \propto n_4}{R} = K\theta'\left(1+\frac{\lambda}{2}\right) \qquad ...(3)$$

where q' is the deflection produced. The constant K can be eliminated from eqns. (2) and (3) *i.e.*, dividing eqn. (2) by (3),

$$\frac{Bn_2 A}{\mu_o n_3 I' \propto n_4} = \frac{\theta}{\theta'}$$

or $$B = \frac{\mu_o n_3 I' \propto n_4}{An_2}\frac{\theta}{\theta} \text{ weber/metre}^2 \qquad ...(4)$$

and $H = n_1 I$ amp/metre

These eqns. give the value of B and H in terms of known quantities. *Experimentally the B-H curve may be plotted either (1) by steps or (2) by reversal.*

(1) *Step method*: In this method the current is increased in steps and ballistic throw is observed. The total current and magnetic field are obtained by adding together the individual values in steps.

(2) *Reversal method*: This method is better because when the current is reversed from + I to – I, the value of B is double as the flux change will be double, hence the ballistic throw q is also doubled. Further, when calibrating B-G, the reversal of current I' doubles the flux density B, consequently the ballistic throw q' is also doubled. Therefore, the value of B does not change as the values of equations (1) and (2) are doubled, hence the ratio of equations (1) and (2) given in equation (4) for B remains unchanged.

DETERMINATION OF SUSCEPTIBILITY C_M AND PERMEABILITY μ

From equation (1),

$$\vec{B} = \mu\vec{H} \qquad ...(1)$$

where $\mu = \mu_o (1 + \chi_m)$

$$\therefore \quad \vec{B} = \mu_o (1 + \chi_m)\vec{H} \quad ...(2)$$

Thus, by knowing $\vec{B}$ and $\vec{H}$, we can calculate the susceptibility cm by using equation (2) and B from equation (1). The $\vec{B}$-$\vec{H}$ curve is shown in Fig. 2.16.

$\vec{B}$-$\vec{H}$ Curve

Since B and H are known, we can also find the value of magnetisation M by the relation,

$$\vec{B} = \mu_o (\vec{H} + \vec{M})$$

Then a curve between $\vec{M}$ and $\vec{H}$ can be drawn it is called $\vec{M}$-$\vec{H}$ curve and is identical to $\vec{B}$-$\vec{H}$ curve.

Variation of Permeability

Although $\mu = \dfrac{B}{H}$, for ferromagnetic materials, μ depends on the magnetising field H. *Various kinds of permeability can be defined from a hysteresis loop.*

1. **Ordinary or normal permeability**. It is the ratio B/H for a point on the normal magnetisation curve, *i.e.*, ratio B/H for the tip of the loop.
2. **Differential permeability**. It is the slope $\dfrac{dB}{dH}$ of the magnetisation curve.

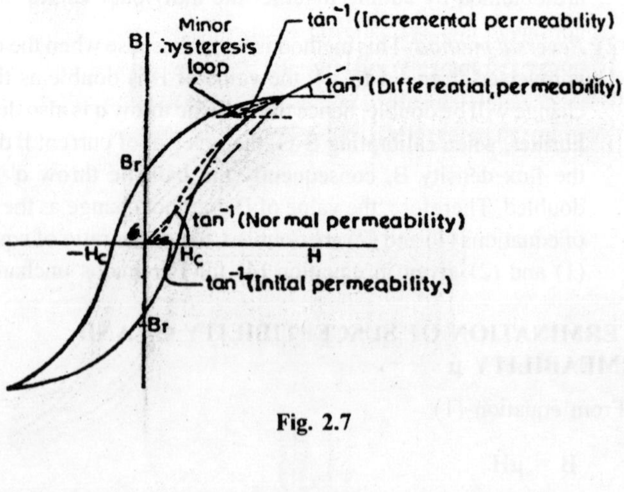

Fig. 2.7

3. **Initial permeability**. The value of $\frac{dB}{dH}$ for H = B = 0 is called the initial permeability.
4. **Incremental permeability**. If a steady value of H is applied to a specimen and then small alternating H is superimposed a minor hysteresis loop is obtained that has its centre near the original point on B-H graph. The slope of the line connecting tips of this minor loop is the incremental permeability. All these are shown in Fig. 2.7.

CYCLE OF MAGNETISATION HYSTERESIS

Let us consider an unmagnetised ferromagnetic substance an iron bar in a magnetising field. The magnetising field is slowly increased as in the case of B-H curve. The magnetisation $\vec{M}$ increases as represented by the part OA. After A any further increase in H would not increase $\vec{M}$. That is the substance reaches the "*magnetic saturation*" $\vec{M}$. In other words, all elementary dipoles have been aligned along the field direction. Now the magnetic field H is gradually decreased to zero. The magnetisation decreases but does not become zero when H = 0. The magnetisation $\vec{M}$ is represented by OB. This value of $\vec{M}$ corresponding to H = 0 is called "*Residual Magnetism*" M_r or "*Retentivity*" or "*Remanence*". The remanence gives the state of permanent magnetisation of the sample. If we consider the B-H curve, the value of B corresponding to H = 0 *i.e.*, after reducing the H from saturation to zero is known as "*remanence induction*" B_r.

If the direction of the field H is now reversed and its magnitude gradually increased, the curve BCD is obtained. The residual magnetism vanishes at C for the value OC of magnetic field in the negative direction. So *OC gives the magnetic field H needed to reduce the magnetisation to zero or in other words this is the magnetising force required to completely demagnetised the sample.* This is called the "*coercive force*" or "*coercivity*" Hc. The three quantities referred above, saturation value, the residual magnetism together with the area of the hysteresis loop are able to explain up to some extent the nature of magnetic material.

Hysteresis

If we further increase H in the negative direction up to the saturation point D symmetrical to A. Now again, all the dipoles are oriented along the field direction but opposite to their direction of orientation at the saturation point A. If we now gradually reduce H to zero and reverse its direction at E, *i.e.,* restore its original direction. Then increasing the field again in steps, we reach back to the point A along the path DEFA. We trace out a path ABCDEFA is called the "hysteresis loop" and is characteristic of the magnetic properties of the material. It is clear from M-H curve shown in Fig. 2.12 that the descending part of the curve always lies above the ascending part. *Hence the magnetisation* $\vec{M}$ *or* $\vec{B}$ *(flux density) will become zero at a 'later point'* of the curve than the field H. It means the *magnetisation lags behind the magnetising field. This lag of magnetisation behind the magnetising field is called hysteresis.*

LANGEVIN THEORY OF DIAMAGNETISM

Let us now examine how the motion of the electron in its orbit is affected by an applied field $\vec{B}$ perpendicular to the plane of the orbit. If A is the area of the orbit, the magnetic flux is BA. According to faraday's law of electro-magnetic induction induced emf ξ is given by,

$$\xi = \frac{d\phi}{dt} = A\frac{dB}{dt} \qquad ...(1)$$

The emf change the motion of the electrons in their orbits. Corresponding to this induced emf, there is an electric field which acts tangentially to the direction of the motion of the electrons. The line integral of this field E around the orbit is equal to the induced emf.

$$\xi = \oint \vec{E}.d\vec{l} = E\oint dl = 2\pi rE$$

or $$E = \frac{\xi}{2\pi r}$$

where r is the radius of the orbit. The electrons are accelerated by this field and experience a force eE. Hence by Newton's law,

$$m_e\frac{dv}{dt} = eE = \frac{e\xi}{2\pi r}$$

and x is given by equation (1), hence

$$m_e\frac{dv}{dt} = \frac{e}{2\pi r}\frac{d\phi}{dt} \qquad ...(2)$$

where me is mass of the electron and v is velocity. Consider, during the time flux changes from 0 to f, the velocity of electron changes from v_o to $v_o + \Delta v$ integrating the equation (2), we get,

$$m_e \int_{v=v_O}^{v=v_O+g\Delta v} \frac{dv}{dt} dt = \frac{e}{2\pi r} \int_0^{\phi} \frac{d\phi}{dt} dt$$

or $$m_e \left|v\right|_{v_O}^{v_O+\Delta v} = \frac{e}{2\pi r} \left|\phi\right|_0^{\phi}$$

or $$\Delta v = \frac{e}{2\pi m_e r}\phi \qquad ...(3)$$

This change Δv gives rise to change in angular velocity of the electron Thus, the change in angular velocity of the electron,

$$\Delta\omega L = \frac{\Delta v}{r} = \frac{e}{2\pi m_e r^2}\phi$$

But ϕ = BA and πr^2 = A, where A is the area of the orbit, we get,

$$\Delta\omega L = \frac{eBA}{2m_e A} = \frac{e}{2m_e}B \qquad ...(4)$$

This change in angular velocity is called "*Larmar angular velocity*" or in terms of frequency "*Larmar frequency*". However, the change in angular velocity acts in sense opposite to that of electron.

Larmar has shown that the effect of the applied field is a "precessional motion" of the orbit without change in its form. The centripetal force necessary to hold the electron in its orbit is given by,

$$F = \frac{m_e v_o^2}{r} = m_e r\omega_o^2$$

where ω_o is the angular frequency of the electron in the orbit without magnetic field. When a magnetic field is applied the angular frequency changes by $\Delta\omega L$. *The condition of stable motion in the same orbit i.e., if electron is still in the orbit of the same radius, its centripetal force must increase correspondingly i.e., the centripetal force.*

$$F' = m_e r\,(\omega_o + \Delta\omega_L)^2$$
$$= m_e r\omega_o^2 + 2m_e r\omega_o\,\Delta\omega_L + m_e r\,\Delta\omega_L^2$$

Assuming that $\Delta\omega_L < \omega_o$ and this corresponds to small value of B thus, term with $\Delta\omega_L^2$ can be neglected. Hence,

$$F' = m_e r\omega_o^2 + 2m_e r\omega_o\,\Delta\omega_L \qquad ...(5)$$

when a magnetic field is applied an additional force acts on the electron.

The magnitude of this force is given by,

$$FB = evB$$

or $\quad FB = er\,(\omega_o + \Delta\omega_L)\,B$ ...(6)

Since $v = \omega_r$ and $\omega = \omega_o + \Delta\omega_L$, FB is perpendicular to both $\vec{v}$ and $\vec{B}$. If electron is moving in anti-clockwise direction and $\vec{B}$ directed out from the orbit, the magnetic force FB will be directed towards the centre of the orbit (*i.e.,* $\vec{B}$ along z-axis $\vec{v}$ along negative y-axis hence $F_B = e\,(\vec{v} \times \vec{B})$ is along + x-axis). If electron is moving in clockwise direction then this force is outward as shown in Fig. 2.18.

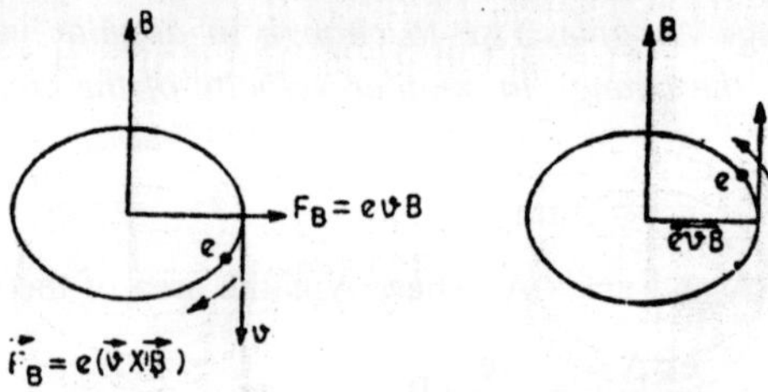

Fig. 2.8

Putting the value of B from equation (5) in eqn. (6),

$$F_B = er\,(\omega_o + \Delta\omega_L)\,\frac{2m_e}{e}\,\Delta\omega_L$$

$$= 2m_e r\omega_o\,\Delta\omega_L + 2m_e r\,\Delta\omega_L{}^2$$

The term $\Delta\omega_L{}^2$ is neglected because of the same reasons. Thus,

$$F_B = 2m_e r\omega_o\,\Delta\omega_L \qquad ...(7)$$

By equation (5), the increases in centripetal force,

$$F' - m_e r\omega_o{}^2 = 2m_e r\omega_o\,\Delta\omega_L \qquad ...(8)$$

The comparison of equations (7) and (8) shows that the increase in centripetal force is completely balanced by the additional inward magnetic force. Hence the centripetal *force does not change by the application of B. Therefore, the radius of the orbit will remain the same.*

A charge e revolving in a circle at the rate of n revolutions per second is "*equivalent*" to a current *ne*. If A is the area of the circle the "*magnetic moment*" of the electron orbit is given by,

$$m = neA, \text{ (since } m = IA)$$

The change in the magnetic moment dm due to change $\Delta\omega_L$ in angular velocity is given by,

$$dm = eA\Delta n. \text{ since } n = \frac{\omega}{2\pi}, \; \Delta n = \frac{\Delta\omega_L}{2\pi}$$

$$= eA\,\frac{\Delta\omega_L}{2\pi}$$

Substituting $\Delta\omega_L$ from eqn. (5), we get,

$$dm = eA\,\frac{eB}{4\pi m_e}$$

$$= \frac{e\pi r^2 eB}{4\pi m_e} \quad (\text{since } A = \pi r^2)$$

$$= \frac{e^2 r^2}{4m_e} B \qquad ...(9)$$

But diamagnetism is an induced effect. This magnetic moment is induced and is always opposite to $\vec{B}$, hence in vector form we can write,

$$d\vec{m} = -\frac{e^2 r^2}{4m_e}\vec{B} \qquad ...(10)$$

We have derived this equation (10) for one electron. It can be extended to an atom having more than one electron. However, in deriving this equation, we have assumed that the orbit is perpendicular to the field. Since the orbit can have any orientation with respect to the field, the radius of the orbit in this equation should be replaced by its projection r_1 on the plane perpendicular to the field, *i.e.*,

$$d\vec{m} = -\frac{e^2 r_1^2}{4m_e}\vec{B} \qquad ...(11)$$

As there are number of randomly oriented orbits within the atom, thus the total magnetic moment induced,

$$\vec{M}_A = -\frac{e^2\vec{B}}{4m_e}\Sigma r_1^2 \qquad ...(12)$$

where summation is taken for all orbits. Suppose (x, y, z) are the coordinates of any point on an orbit of radius r, then,

$$r^2 = x^2 + y^2 + z^2$$

If we choose z-axis in the plane perpendicular to the applied magnetic field, then

$$r_1^2 = x^2 + y^2$$

Let the mean value of the radii of all orbits be r,

$$r^2 = x^2 + y^2 + z^2$$

Fro spherically symmetrical atom,

$$x^2 = y^2 = z^2 = \frac{1}{2} r^2$$

hence, $$r_1^2 = x^2 + y^2 = \frac{2}{3} r^2$$

Substituting the value of r_1^2 in eqn. (12), we get,

$$\vec{M}_A = -\frac{e^2 \vec{B}}{6m_e} \Sigma r^2 \quad ...(13)$$

For a substance containing n molecules per unit volume, the total magnetic moment induced per unit volume of the substance,

$$\vec{M}_v = -\frac{e^2 n\vec{B}}{6m_e} \Sigma r^2 \quad ...(14)$$

By equation (14), the susceptibility,

$$\chi_m = \frac{\text{Magnetic moment}}{H}$$

The diamagnetic susceptibility per atom is called Atomic susceptibility denoted by χ_{mA}. Then,

$$\chi_{mA} = \frac{\text{Induced magnetic moment per atom } M_A}{H}$$

Substituting MA from equation (14), we get,

$$\chi_{mA} = -\frac{e^2 B}{6m_e H} \Sigma r^2$$

For diamagnetic substance, we have linear relation,

$$B = \mu H$$

Hence, $$\chi_{mA} = -\frac{\mu e^2}{6m_e} \Sigma r^2 \quad ...(15)$$

Similarly, the volume susceptibility,

$$\chi_{mv} = -\frac{n\mu e^2}{6m_e} \Sigma r^2 \quad ...(16)$$

The scrutiny of the above results leads to the following conclusions:

1. The expression for susceptibility is independent of temperature.
2. The diamagnetism arises from the effect of magnetic field on the motion of electrons in the orbit and every substance having moving electrons in the orbit. Hence, every substance possesses diamagnetism, but diamagnetism is a feeble effect and overridden by the paramagnetism and ferromagnetism producing a flux in the same direction as that externally applied field.
3. The negative sign indicates that diamagnetism is an induced opposing effect, that disappears as soon as the field is removed.

ATOMIC CURRENTS OR AMPERIAN CURRENTS

A careful study of the various magnetic phenomena associated with matter indicates that magnetism is an atomic or molecular property and that there is no intrinsic magnetic substance. In the light of this, the hypothesis is adopted that all magnetic properties of matter can be explained in terms of the motions of the electrons and positive charges associated with its constituent atoms and molecules.

All matter is made of atoms each of which consists of electrons those move in the orbits around the nucleus and rotate or spin about their own axe. These circulating electrons constitute a current without a charge transport, we call this as "Atomic currents" and is different from true-current which corresponds to charge transport. Such circulating currents were first postulated by Ampere, and are often called Amperian currents.

GAUSS'S LAW OF MAGNETISM

The relation div $\vec{B} = 0$ (everywhere) is mathematically representing the fact that lines of $\vec{B}$ are closed on themselves, whenever the divergence of vector point function is zero, it means no sources or sinks exist in the region over which the divergence is zero.

In the same way Gauss's law is a formal way of stating a fundamental fact of magnetism, *i.e.*, the isolated magnetic pole does not exist, the law (we have already discussed) states that the magnetic flux through any closed surface is zero, *i.e.*,

$$\phi_B = \oint \vec{B}.d\vec{S} = 0 \quad ...(1)$$

where integral is to be taken for entire surface. In contrast for electric field, we have

$$\varepsilon_o \oint \vec{B}.d\vec{S} = q \text{ or div } \vec{E} = \frac{\rho}{\varepsilon_o} \quad ...(2)$$

By the comparison of the two equations A and B it is evident that in magnetism there is no counterpart to the free charge q.

MAGNETIC MOMENT AND ANGULAR MOMENTUM (BOHRMAGNETON)

An electron is circulating in a circular orbit of radius r is equivalent to current loop with the current flowing in a direction opposite to v (the velocity of electron). Let us examine this. An electron is moving in a orbit of radius r with velocity v The motion of electron in orbit is governed by Bohr's rule. The force between an electron in the orbit of radius r and the nucleus of hydrogen atom is given by Coulomb's law.

$$F_e = \frac{1}{4\pi\varepsilon_o}\frac{e^2}{r^2}$$

and electron is bound by this force with nucleus, but it moves in a circular orbit of radius r. The centripetal force $Fc = \frac{m_e v^2}{r}$ holding the electron in an orbit r from the nucleus is provided by the electrostatic force $F_e = \frac{1}{4\pi\varepsilon_o}\frac{e^2}{r^2}$ between them, and the condition for orbit stability is $F_c = F_e$ *i.e.,*

$$\frac{1}{4\pi\varepsilon_o}\frac{e^2}{r^2} = \frac{m_e v^2}{r} \quad ...(1)$$

or

$$v = \sqrt{\frac{e^2}{4\pi\varepsilon_o m_e r}}$$

The angular velocity ω is given by

$$\omega = \frac{v}{r} = \sqrt{\frac{e^2}{4\pi\varepsilon_o m_e r^3}} \quad ...(2)$$

The orbital angular momentum is

$$L_l = m_e vr$$

Substituting v from eqn. (1), we get,

$$L_l = \sqrt{\frac{e^2 m^2 r}{4\pi\varepsilon_o}} \qquad ...(3)$$

Its direction is determined by cross-product rule, as $\vec{L}_l = \vec{r} \times \vec{p}$, where $\vec{P} = m\vec{v}$. Thus L_l is perpendicular to $\vec{r}$ and $\vec{v}$, if $\vec{r}$ is along x-axis, $\vec{v}$ is along y-axis then $\vec{L}_l$ is along z-axis. $\vec{L}_l$ is directed perpendicular to the plane of the orbit as shown upward.

Now the rotating electron constitute the current, and the current is determined by the rate at which the charge passes the given point. If v is the velocity of the electron it will make $\frac{v}{2\pi r}$ revolutions per second and is equivalent to a current given by

$$i = \frac{ev}{2\pi r}$$

$$= e\frac{\omega}{2\pi}, \text{ substituting w from above Eqn. (2)}$$

$$= \frac{e}{2\pi}\sqrt{\frac{e^2}{4\pi\varepsilon_o m_e r^3}}$$

$$= \sqrt{\frac{e^4}{16\pi^2\varepsilon_o m_e r^3}}$$

The close orbit will behave as a magnetic dipole and its magnetic moment is given by

$$m_l = iA$$

$$= \sqrt{\frac{e^4}{16\pi^2\varepsilon_o m_e r^3}} \cdot \pi r^2$$

$$= \frac{e^2}{4}\sqrt{\frac{r}{\pi\varepsilon_o m_e}} \qquad ...(4)$$

Its direction is given by right hand curl rule. Let the fingers of the RH curl around the loop in the direction of the current, the extended right thumb will then point in the direction of $\vec{m}_l$. However, by relation j = nev and i = j/A the direction of current is in the direction of v. Thus ml is also directed perpendicular to the plane of the orbit, so Ll and ml are in the same direction. The ratio

$$\frac{m_l}{L_l} = \frac{e^2}{4}\sqrt{\frac{r}{\pi\varepsilon_o m_e}}\sqrt{\frac{e^2 m_e r}{4\pi\varepsilon_o}} \quad \text{(substituting from eqns. 3 and 4)}$$

$$= e/2\ me \qquad ...(5)$$

which shows that the orbital magnetic moment of an electron is proportional to its orbital angular momentum.

ORBITAL GYROMAGNETIC RATIO

The ratio of the magnetic dipole moment of the electron due to its orbital motion and the angular momentum of the orbital motion is called orbital gyromagnetic ratio, or sometimes orbital magneto mechanical ratio of an electron. Thus,

$$\frac{\text{magnetic moment}}{\text{Angular momentum}} = \frac{e}{2m_e}$$

From Bohr's postulates, the electron chooses such orbits, for which the angular momentum L_l is an integral of $h/2\pi$ where h is Planck's constant *i.e.,*

$$m_l = \frac{nh}{2\pi}\frac{e}{2m_e} \qquad ...(1)$$

$$= n\left(\frac{eh}{4\pi m_e}\right)$$

where n = 1, 2, 3, ... etc. ...(2)

All the quantities in $\frac{eh}{4\pi m_e}$ are constants and together form a unit of magnetic moment called the Bohr magneton, *i.e.,*

$$\text{Bohr magneton} = \frac{eh}{4\pi m_e} \qquad ...(3)$$

substituting e, h and m, we get,

Bohr magneton = 9.1×10^{-24} amp.m^2

And for first orbit n = 1, from equation (3), the magnetic moment is one Bohr magneton.

ELECTRON SPIN

S.A. Goudsmith and G.E. Uhlenbeck proposed in 1925 that an electron while moving in circular orbit round the nucleus also rotates about its own axis just as the earth rotates about its axis. In other words the electron has a spin motion over and above the orbital motion. The rotation involves angular momentum, and because the electron is negatively charged it has a magnetic moment Ms. opposite in direction to its angular momentum vector L_S as shown in Fig. 2.9.

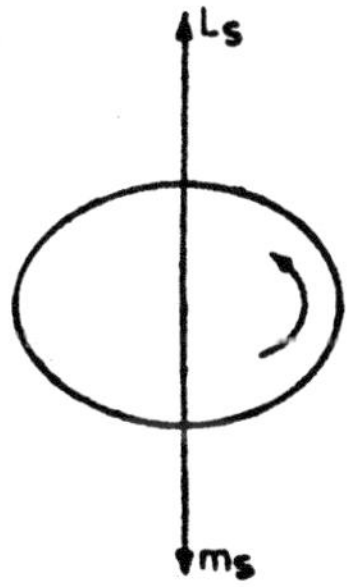

Fig. 2.9

With thc idea of spin the electron possess an intrinsic angular momentum independent of any orbital angular momentum it might have and associated with this angular momentum is a certain magnetic moment m_S. The detailed theory which is purely quantum mechanical is beyond the scope here, however, the ratio of magnetic moment m_S to the angular momentum L_S for the electron spin is twice as that for its orbital motion, *i.e.*,

$$\frac{m_S}{L_S} = \frac{e}{m_e}, \text{ [from equation (1)]} \qquad ...(1)$$

The magnitude of L_S always comes out to be $\frac{1}{2}\left(\frac{h}{2\pi}\right)$ where h is Planck's constant.

In an atom there are several electrons each one has angular momentum as well as magnetic moment for both orbital and spin motion. Thus, an atom have a total angular momentum and a total magnetic moment. For an atom as a whole the ratio of magnetic moment to angular momentum on the basis of quantum physics is e/m, or e/2m, or in between these two because there is a combination of the contributions from the orbits and spins,

$$m_A = g\left(\frac{e}{2m_e}\right)L \qquad ...(2)$$

where m_A is total magnetic moment and L is total angular momentum and g is called Lande g-factor which is the characteristic of the atom. The value of g is one for a pure orbital moment, two for a pure spin moment and for an atom g lies between one and two because electrons in atoms have both orbital and spin motion.

In nuclei there are protons and neutrons and they also more around in orbits and spin like electrons. The magnetic moment is parallel to the angular momentum. Therefore for nuclei, we have,

$$m_{nuclei} = g\left(\frac{q_e}{2m_p}\right)L \quad ...(3)$$

where q_e is charge, m_p is mass of proton and g is called the nuclear g factor, q is a number near one to be determined for each nucleus. However the spin magnetic moment of the proton does not have a g factor of two, *i.e.,* the value for the electron. For proton g = 2.79. The neutron although neutral also has a spin magnetic moment, The neutron is not exactly neutron in the magnetic sense.

ENERGY LOSS PER CYCLE OF MAGNETISATION

The process of magnetisation of a ferromagnetic substance through a cycle is involves the expenditure of the energy. This energy is spent in aligning the elementary dipoles in the direction of magnetising field. If we gradually reduce the magnetising field to zero, the magnetisation does not fall to zero and magnetic field has to be applied in the opposite direction to completely demagnetise the sample.

Thus energy spent in magnetising a specimen is not completely recoverable and there is a loss of energy in taking a sample through a cycle of magnetisation and it appears in the form of heat. This loss of energy is called hysteresis loss.

Hysteresis Loss from B-H Curve

Let us consider the sample of ferromagnetic material in the shape of toroidal solenoid. The magnetising field H is obtained by modified form of Ampere's law and is given in equation (1) *i.e.,* $\oint \vec{H}.d\vec{l} = I$, $H_l = NI$ or $H = \frac{NI}{l}$ where N are total number of turns over toroidal winding and l is length of toroid

thus $H = n_1 I$

where n1 are the number of turns per unit length, then we get,

$$\vec{H} = \frac{NI}{l} \qquad ...(1)$$

Let us calculate the amount of work that has to be done in establishing a current I in the toroidal coil by which magnetising field H is obtained and the substance is magnetised giving rise to the magnetic flux density (or induction) $\vec{B}$. If A is the area of toroidal ring, the flux inside the ring is,

$$\phi = NAB$$

Now as current I increases, the magnetic flux also increases and according to Faraday's law an induced emf develops in the coil, the magnitude is given by,

$$\xi = \frac{d\phi}{dt} = NA\frac{dB}{dt}$$

The rate of doing work,

$$P = \xi I = NAI.\frac{dB}{dt} \qquad ...(2)$$

Substituting I from equation (2),

$$P = HA_l\frac{dB}{dt}$$

The product A_l is the volume of the sample *i.e.*, ring. Hence the rate of doing work per unit volume,

$$P = H\frac{dB}{dt}$$

This gives the rate at which energy is transferred per unit volumes to the sample in magnetising it is,

$$\frac{dw}{dt} = P = H\frac{dB}{dt}$$

Therefore the energy transferred in time dt per unit volume,

$$dw = HdB$$

The energy transferred to the sample per unit volume per cycle

$$= \oint HdB \qquad ...(3)$$

Now be careful to the fact that through it has been deduced for an iron ring this expression is true in general.

Let us evaluate the integral for ferromagnetic materials. Ferromagnetic materials are "inherently irreversible" as they retain magnetisation even if the magnetising field is reduced to zero. Fig. 2.10 represents a part of B-H loop. For ferromagnetic materials the small change in H produces a large change in $\vec{B}$ (or $\vec{M}$). But for the change dH in H, the corresponding change dB in B will be very small. Therefore, the value of H may be taken to be same for the points a and b. The product HdB is the energy transferred per unit volume to the sample for a change dB in B is equal to the area ab × cb which is area abcd shown shaded. Therefore, the energy transferred for the path OA per unit volume

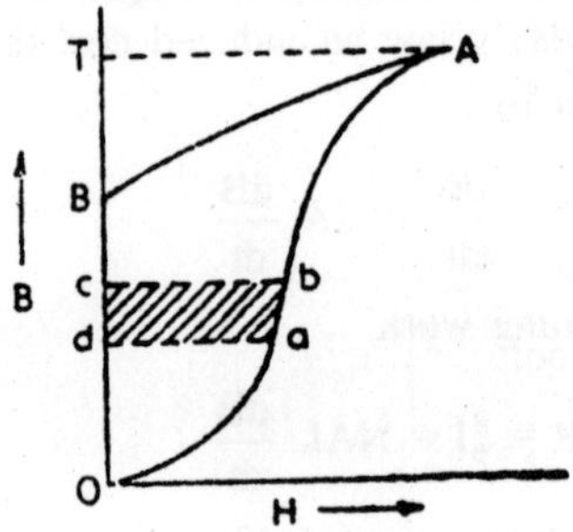

Fig. 2.10

$$= \int_{OA} HdB$$

= area OATO

Now H is gradually decreased reduced to zero as the process is irreversible. The trace does not follow AO but AB. Hence, the energy recoverable during demagnetisation

= area BATB

Therefore the energy Energy transferred –Energy recovered

not recoverable = during magnetisa-du.. ıg demagneti-

i.e., energy losstion sation

= area OATO – area BATB

= area OABO

If we consider the hysteresis loop, then the energy not recoverable, *i.e.*, energy loss,

$$\oint HdB = \text{area ABCDEFA}$$

$$= \text{area B-H loop} \qquad ...(4)$$

The energy loss is converted into heat.

Energy density

The energy required to magnetise the sample per unit volume per cycle is given by eqn. (4), *i.e.*,

$$= \oint HdB$$

It is assumed that the relation B = μH is valid over the entire range of B with $\mu = \mu_o \mu_r$ then energy stored in the sample if B changes from 0 to B.

$$\int_0^B HdB = \int_0^B \frac{B}{\mu} dB = \frac{B^2}{2\mu} = \frac{\mu H^2}{2} = \frac{BH}{2} \text{ Joules/metre}_2$$

Magnetic energy per unit volume is called,

Magnetic energy density

$$= \frac{1}{2}\frac{B^2}{\mu} \text{ Joules/metre}^2$$

$$= \frac{1}{2}\mu \text{ } H_2 \text{ Joules/meter}^2 \qquad ...(5)$$

$$= \frac{1}{2} BH \text{ Joules/metre}^2$$

where the permeability of the medium is μ. In case the medium is vacuum (or air), $\mu_r = 1$, $\mu = \mu_o$.

$$\text{Energy density} = \frac{1}{2}\frac{B^2}{\mu_o}$$

which is same as derived earlier in equation (5). This expression for energy is analogous to the electrostatic energy expression *i.e.*,

$$\frac{1}{2} \epsilon_o E^2 \text{ or } \frac{1}{2}\varepsilon E^2$$

Hysteresis Loss from MH. Cycle

Suppose the behaviour of the ferromagnetic substance is represented by M-H curve. Using the relation,

$$\vec{B} = \mu_o (\vec{H} + \vec{M})$$

or $$dB = \mu_o (dH + dM)$$

Integrating for whole hysteresis loop, we have after multiplying by H,

$$\oint HdB = \mu_0 \oint HdH + \mu_0 \oint HdM$$

If H is plotted against dH we will get a straight line, the integral $\oint HdH$ *represents the area enclosed by a straight line which is zero.* So we have,

$$\oint HdB = \mu_0 \oint HdM$$

$\oint HdB$ represents the energy loss per unit volume per cycle and equal to area of $\vec{B}$-$\vec{H}$ loop. Similarly $\oint HdM$ equals to area of $\vec{M}$-$\vec{H}$ loop or curve.

Therefore the energy loss = μ_0 area of M-H loop ...(6)

This is called Warburg's law.

Thus, *the energy loss per unit volume of the material per cycle is equal to* μ_0 times the area of M-H loop and is dissipated in the form of heat.

DEMAGNETISATION

Ferromagnetic materials are not fully demagnetised by removing the magnetising field. They retain magnetisation corresponding to H = 0. The capacity to retain magnetisation depends on the degree up to which the substance is magnetised. It is maximum corresponding to saturation value of magnetisation. The portion of a saturation hysteresis loop for which M is positive H is negative is called "demagnetisation curve".

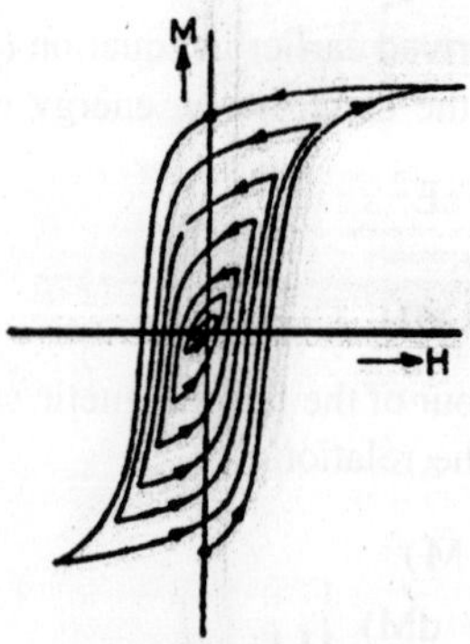

Fig. 2.11

The retentivity is different for different values of H below saturation value. In fact retentivity decreases with decreasing H below saturation value. Hence it is possible to reduce retentivity to zero value by gradually decreasing H, through repeated cycles as shown in Fig. 2.11. However the demagnetisation is also possible by heating the substance above the Curie temperature. But this method causes the permanent change in the magnetic behaviour of the substance.

Soft and Hard Magnetic Materials

The shape of hysteresis loop is a characteristics of the magnetic material under test. Fig. 2.12 shows M-H curves for iron and steel. We can draw the following conclusions from the curves.

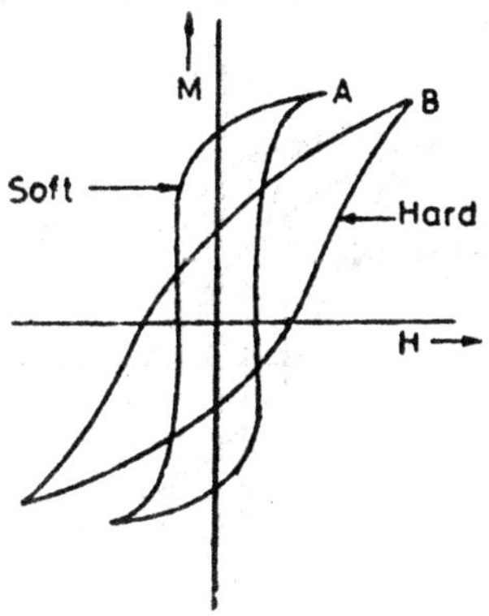

Fig. 2.12

Coercivity

Substances with large value of coercive force are said to be magnetically hard, whereas those small coercive force are called soft.

Retentivity

The retentivity is greater for soft materials than for hard material.

Susceptibility

It is greater for soft materials than for hard materials.

Permeability

It is greater for soft materials than for hard materials.

Hysteresis Loss

The area of the hysteresis loop which measures the energy loss per unit volume per cycle is less for soft materials than for hard materials.

USES OF HYSTERESIS CURVE CHOICE OF MATERIALS

The hysteresis curve and its properties, saturation, retentivity, coercivity and hysteresis loss provide very useful information for choosing the material for practical uses in industry

Permanent Magnets

For permanent magnets the material should have

(1) Large coercive force because we want that its magnetism should be permanent and not easily destroyed by magnetic fields, mechanical ill-treatment or by temperature change, and

(2) High residual magnetism so that the magnet may be strong.

These requirements are satisfied by steel. Though the retentivity of iron is greater than steel but iron has very small value of coercivity, *hence steel is preferred for permanent magnets.* The fact that hysteresis loss is also less for soft iron is immaterial because permanent magnets are never put to cyclic change of magnetisation. The materials used are cobalt-steel, tungsten steel and other number of alloys. Very recently an alloy of iron, cobalt and vanadium called vicalloy having highest coercivity has been developed.

Electromagnets

The materials for the construction of core of electromagnets should have

1. maximum induction B with comparatively small magnetising field H.
2. high initial permeability.
3. low hysteresis loss.

Considering these facts the *soft iron is an ideal material for constructing an electromagnet.*

Transformer Cores, telephone diaphragms, Armature of dynamos and motors and cores of chokes

When these are used practically the material is subjected to cyclic changes. The chief requirements are that the material must have,

1. high initial permeability in order to obtain large B for low value of H since ($B = \mu H$).
2. low hysteresis loss which will produce a small heating effect hence winding is secure of any breakdown of insulation, and
3. high specific resistance to reduce eddy current losses.

Soft iron is better than steel for these purposes. It has initial permeability about 250, initial permeability can be increased by adding 4 per cent silicon and alloy is called transformer steel which is a very good material for cores of transformers. There are some other alloys containing nickel iron called Term alloys, Mumetal (Ni 76%, Cu 5%, Fe 17.5%, Cr 1.5%) having initial permeability between 10,000 and 100,000. An alloy called radio-metal (N 48%, Fe 48.5%, Cu 3%, Mn 0.5%) has an initial permeability between 2,000 to 15,000.

Magnetomotive Force and Magnetic Circuits

We are familiar with an electrical circuit and Ohm's law which connects the emf with the current in the circuit. There is flow of electrons which constitutes the electric current. The line integral of electric field $\vec{E}$ *i.e.*, $\oint \vec{E}.d\vec{l}$ over a circuital path is the electromotive force and is equal to the work done in carrying a unit positive charge once round the circuit. But in magnetism, since there is nothing like a separate north pole and a thing as magnetic currents is meaningless. However, the magnetic lines of flux form closed loops. Such closed paths are known as magnetic circuit and we have Ampere circuital law that gives the line integral of magnetic field H *i.e.*,

$$\oint \vec{H}.d\vec{l} = \text{enclosed current}$$

Hence by analogy we can say that the line integral $\oint H.d\vec{l}$ *is equal to the work done in currying unit north pole once round the close path if north pole is free to do so and termed as mmf (magnetomotive force)*

Hence, from Ampere's law,

$$\oint \vec{H}.d\vec{l} = H\oint dl = I$$

Here $\vec{H}.d\vec{l}$ = Hdl as H is everywhere parallel to the path $d\vec{l}$ and $\vec{H}.\oint dl$ = length of the path.

Let us consider the case of toroidal solenoid of N turns. The total current enclosed is NI, the magnetic field H inside is given by,

$$H = \frac{NI}{l} \text{ amp/metre}$$

The flux density inside the right is,

$$B = \mu H$$

$$= \frac{\mu HI}{l}$$

Total magnetic flux $\phi_B = \int_S \vec{B}.d\vec{S}$

$$\phi_B = BA = \frac{\mu NIA}{l}$$

where A is area of cross-section of ring. Therefore,

$$NI = \frac{l}{\mu A}\phi_B \quad ...(1)$$

Returning now to the consideration of electric field, we know that emf,

$$\xi = \oint \vec{E}.d\vec{l}$$

In a magnetic field we may write an "analogous relation", based on Ampere's law, that when current is enclosed by the path of integration,

$$\oint \vec{H}.d\vec{l} = NI = \xi_m \quad ...(2)$$

where the quantity ξ_m, called "magneto motance", or magnetomotive force" *i.e.,* mmf is equal to the current enclosed. The product NI is expressed in ampere turns.

Unit of mmf

From above equation (2), the mmf has the same unit as NI *i.e.,* ampere turns. In N = 1, then simply ampere. Comparing equation (1) and (2),

$$\text{mmf} = NI = \phi_B \frac{l}{\mu A} \quad ...(3)$$

In an electrical circuit we have similar relation, namely,

emf = current × resistance

(by Kirchhoff's law the total emf in the circuit is equal to the total IR drop). We can write,

$$\text{Current} = \frac{\text{emf}}{\text{Resistance}}, \text{ i.e., } I = \frac{\xi}{R} \quad ...(4)$$

and from equation (2.59),

$$\text{Flux} = \frac{\text{mmf}}{l/\mu A}, \textit{ i.e., } \phi_B = \frac{\xi m}{R_e} \qquad ...(5)$$

By analogy we define a quantity called the "reluctance" R_e which is considered to oppose the setting up of magnetic flux in magnetic circuit as resistance opposes the setting up electric current. R_e is expressed in amp. turns/weber which is reciprocal of inductance.

Therefore the unit for reluctance are:

$$\frac{\text{amp. turns}}{\text{Weber}} = \frac{1}{\text{Henries}}$$

Let us consider the reluctance of portion of a magnetic circuit, since in an electric circuit we have Ohm's law,

$$R = \frac{V}{I}$$

where V is potential difference between the points.

In the analogous magnetic case reluctance R_e between two points,

$$R_e = \frac{V_m}{\phi_B} \qquad ...(6)$$

where V_m = magnetic potential difference between the points, and

$$V_m = \int_1^2 \vec{H}.d\vec{l}, \text{ just as } V = \int_1^2 \vec{E}.d\vec{l}$$

Thus, we call equation (6) as *"Ohm's law" in magnetism.*

Permeance

The reciprocal of reluctance is known as permeance

$$\frac{1}{R_e} = \text{Permeance, its unit is } \frac{\text{Weber}}{\text{amp.}} = \text{henrys}$$

and ϕ_B = mmf × permeance

$$\text{Reluctance} = \frac{l}{\mu A}$$

$$\text{and Resistance} = r\frac{l}{A} = \frac{1}{\sigma}\frac{l}{A}$$

where σ is electrical conductivity, μ plays the same part in magnetic circuit as σ does in electrical circuit.

Combination of reluctances

If a magnetic circuit consists of different materials, with different length and cross-section. The combined reluctance is obtained by adding individual reluctances, as we do for resistances in series. Thus in a magnetic circuit made up of different materials of permeabilities μ_1, μ_2... of lengths l_1, l_2... and of cross-section A_1, A_2..., the total reluctance is given by

$$R_e = \frac{l_1}{\mu_1 A_1} + \frac{l_2}{\mu_2 A_2}$$

or $$R_e = \Sigma \frac{l}{\mu A} \quad ...(7)$$

Points of difference in two circuits

In spite of the above analogy in two circuits with regard to various terms, they differ on the following points:–

1. In an electric circuit, the current represents the actual flow of electrons or ions whereas in magnetic circuit there is no such flow because free poles do not exist.
2. In an electric circuit, energy is required to create as well as maintain the electric current. But in magnetic circuit the energy is required only for creation of the flux.
3. Though apparently μ corresponds to σ, the specific conductivity but actually μ in magnetic process corresponds to ε which is the absolute permittivity of the medium in electrical process.
4. An important point of difference is, that the resistance is independent of the current in the circuit. As μ is not constant, the reluctance of magnetic circuit depends upon the flux.

Magnetic Circuit with Air Gap

Consider an iron ring of permeability μ, having mean circumference $l(= 2\pi r)$ and area of cross-section A and a narrow air gap of thickness d. The field H_g in the gap, $H_g = \frac{B}{\mu_0}$ while the field H_i in the iron, $H_i = \frac{B}{\mu}$

$$H_i = \frac{B}{\mu} = \frac{B}{\mu_r \mu_0} = \frac{H_g}{\mu_r}$$

or $$\frac{H_g}{H_i} = \mu_r \quad ...(8)$$

The reluctance of air gap $= \dfrac{d}{\mu_0 A}$

The reluctance of iron $= \dfrac{l-d}{\mu A}$

Total reluctance $= \dfrac{d}{\mu_0 A} + \dfrac{l-d}{\mu A}$

The magnetic flux,

$$\phi_B = \frac{NI}{\dfrac{d}{\mu_0 A} + \dfrac{l-d}{\mu A}} \quad ...(9)$$

In the absence of air gap,

$$\phi_B = \frac{NI}{l/\mu A} \quad ...(10)$$

The comparison of equations (8) and (10) shows that the magnetic flux in the circuit with air gap is less than that in the circuit without any air gap. *Thus with air gap flux is considerably reduced.*

Magnetic Circuit of an Electromagnet

Fig. 2.13 shows an electromagnet with an air gap and with different ferromagnetic materials of various shapes and lengths. With the idea of magnetic circuit it is easy to calculate the flux by knowing the total reluctance of the circuit. The total reluctance is given by,

$$\frac{l_1}{\mu_1 A_1} + \frac{2l_2}{\mu_2 A_2} + \frac{l_3}{\mu_3 A_3}$$

where A_1, A_2 and A_3 are areas of cross-section for three parts of the electromagnet.

$$\text{Flux } \phi_B = \frac{\text{mmf}}{\dfrac{l_1}{\mu_1 A_1} + \dfrac{2l_2}{\mu_2 A_2} + \dfrac{l_3}{\mu_3 A}} = \frac{\text{mmf}}{\Sigma \dfrac{l}{\mu A}} \quad ...(11)$$

As μ_2 corresponds to air gap $\mu_3 = \mu_0$, because μ_r the relative permeability for air is one. However, the calculation is only an approximation because leakage will take place in air gap. If air gap is very thin leakage is negligible and calculations are more reliable. *The small air gap l_3 increases the reluctance of the circuit by a large amount.*

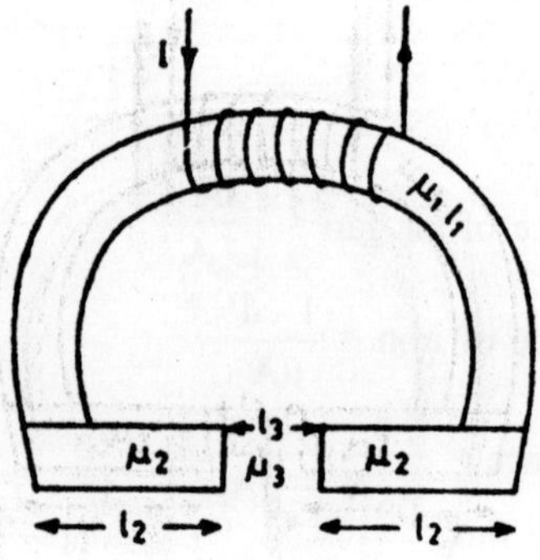

Fig. 2.13

Calculation of Lux due to a Close Ring

Such a ring is shown in Fig. 2.14. With different areas of cross-sections, A_1, A_2, A_3, ... at different points the corresponding flux densities are B_1, B_2, B_3. Assuming that there is no flux leakage and using Gauss's law for magnetism, *i.e.*,

$$\oint_S \vec{B} \cdot d\vec{S} = 0$$

i.e., the number of lines of $\vec{B}$ leaving any volume enclosed by a surface will be the same as that entering it. Hence, we get,

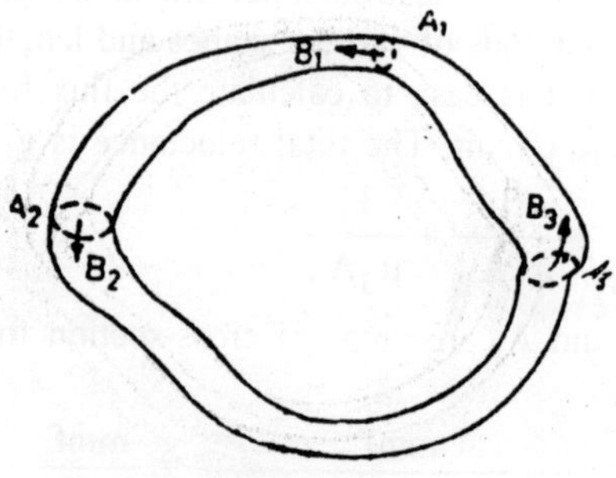

Fig. 2.14

$$\phi_B = B_1A_1 = B_2A_2 = B_3A_3 \ldots = B_nA_n$$

ATOMIC MODEL AND MAGNETISM

An electron revolving in its orbit around the nucleus of an atom forms a tiny electric current loop. Since a current loop has a magnetic field and *all atoms have revolving electrons we might suppose that all substances would exhibit magnetic effects.* It is reasonable to imagine that magnetic field associated with a piece of iron is due to tiny electric currents within the iron itself. Such electric currents are known as Amperian currents. It can be shown that "magnetic moment" of a current loop is m = Ia where a is the area of the orbit. This is often referred to as the "orbital magnetic moment".

Another feature of the atomic model, which is based on experimental and theoretical results is that the election in the atom also rotates about an axis through the electron (for detail see art. 4.5 electron spin). The electron spin is considered as the equivalent of a tiny current (amperian current) consequently there is an associated magnetic moment with the electron spin. This is referred to as the "spin magnetic moment". The resultant magnetic moment of the atom is obtained by the proper vector summation of the orbital and spin magnetic moments. These are combined according to quantum theory rules, which will not be discussed here. It is important to note, however, that *the magnetic properties of bulk matter can be described in terms of the orbital and spin motions of the atoms which make up the matter.*

Classification of Magnetic Material

In many atoms the electrons are completely paired, that its, for each electron spinning in one direction, there is an electron spinning in the opposite direction. And the same situation exists in regard to the orbital motion of the electrons. Thus, the net current circulating about any given axis is zero and there are no Amperian current in such substances, consequently the resultant magnetic moment of the atom is also zero. *Such substances show very "weak magnetic effects" that they are called "non-magnetic". However, vacuum is the only truly non-magnetic medium.* All materials show some magnetic defects with the exception of the ferromagnetic group these effects are weak. It has been established both theoretically and experimentally that all matter may be classified into three groups, according to its fundamentally different behaviour under the action of a magnetic field. These are classified as (1) diamagnetic, (2) paramagnetic and (3) ferromagnetic

Diamagnetic Substances

Diamagnetic substances are those which, when placed in a magnetic field, *become weakly magnetised. The resultant magnetisation is opposed to the applied field.* The magnetism which is thus established, in general, varies directly with the impressed field and is independent of the absolute temperature. A diamagnetic substance contains no dipoles, but only dipoles that are induced by an external field.

Paramagnetic Substances

Paramagnetic substances are those which, when placed in a magnetic field. *Become weakly magnetised in the direction of the applied field.* The magnetism thus established, in general, is proportional to the impressed field and varies inversely with the absolute temperature. A paramagnetic substance contains permanent dipoles.

Ferromagnetic Substances

Ferromagnetic substances are those which, when placed in a magnetic field, *become strongly magnetised in the direction of the applied field,* and which also exhibit the phenomena of retentivity and hysteresis. The magnetism whitish thus established in general, is proportional to the applied field and varies with the quantity T – θ, where T is the absolute temperature and θ is the critical temperature and is characteristics of the material. The variation is known as Curie Weiss law. *Ferromagnetic materials are sometimes classed as strongly paramagnetic.* All ferromagnetic substances are paramagnetic substances but all paramagnetic substances are not ferromagnetic substances. The ferromagnetism disappears are not ferromagnetic substances. The ferromagnetism disappears if the temperature is increased above a critical value called Curie temperature and substance becomes paramagnetic.

LANGEVIN'S THEORY OF PARAMAGNETISM

As we have pointed earlier that the molecules or atoms with permanent magnetic dipole moment are termed paramagnetic. For such molecules there is atomic (or amperian) current which gives rise to the magnetic dipoles having magnetic dipole moment. They have amperian current, because the orbit of the electrons are so oriented that the amperian currents due to their orbital motion as well as spin do not cancel out completely. This happens with all the atoms and molecules having odd number of electrons for them the total spin is not zero. In most bulk materials, there is a net magnetic moment only if there are atoms present whose inner electron shell is no filled.

Langevin theory of paramagnetism is just like the Langevin theory of dielectrics (polar-molecules). Suppose each atom or molecule possesses a permanent magnetic moment m. When an external magnetic field $\vec{B}$ is applied, the *dipoles align along the direction of applied field.* Let the axis of dipole be making an angle θ with the direction of applied field B. The magnetic potential energy is given by,

$$U = -\vec{m}.\vec{B}$$

$$= -mB\cos\theta$$

However, the alignment of dipoles will be counteracted by the collisions of thermal motion between the molecules which knock the axes out of line. As a result, there will be an equilibrium distribution of the axes with reference to the direction of the field, which can be determined with the help of the kinetic theory of gases as follows.

According to the kinetic theory principle, the number of molecules distributed with the axis of the dipole pointing in all directions within a solid angle Ω at an angle θ with a given line of reference is proportional to dΩ. And on the basis of Boltzmann theorem, the number of molecules whose potential energy is v is proportional to $e^{-U/KT}$. Considering unit volume of the gas, if dN out of the total number N of the dipoles in it have their magnetic axes inclined at the angle π with the field direction, then, applying the principles of kinetic theory states above, we get,

$$dN \propto e^{-U/KT}\, d\Omega$$

$$= C\, e^{-U/KT}\, d\Omega$$

where C is constant depending on the number of molecules considered, U is potential energy, K is Boltzmann constant and T is absolute temperature. We have calculated that

$$d\Omega = 2\pi \sin\theta\, d\theta \text{ and } U = -m\,B\cos\theta$$

After substitution, we get

$$dN = C\exp\left(\frac{mB\cos\theta}{KT}\right) 2\pi\sin\theta\, d\theta$$

Putting, $$\frac{mB}{KT} = x \qquad ...(1)$$

The component of the magnetic moment of one molecule in the direction of field is m cos θ. Therefore, the total magnetic moment of dN molecule, along the field direction is,

$$dM = dN\, m\cos\theta$$

$$= C \text{ excosq } 2\pi \sin\theta \text{ m}' \cos\theta \, d\theta$$

$$= 2\pi \text{ C m } \sin'\theta \cos\theta \, e^{x\cos\theta} \, d\theta \qquad \text{...(2)}$$

Since the possible orientation are contained between 1 and π, lance the total magnetic moment M of all molecules in unit volume in the direction of field obtained by integrating the above expression,

$$M = \int_0^\pi 2\pi \, C \, m \sin\theta \cos\theta \, e^{x\cos\theta} d\theta$$

$$= 2\pi \text{ C m} \int_0^\pi e^{x\cos\theta} \sin\theta \cos\theta \, d\theta \qquad \text{...(3)}$$

The total number of molecules N in unit volume can be found by integrating dN,

$$N = \int_0^\pi dN = \int_0^\pi 2\pi \, C \, e^{x\cos\theta} \sin\theta \, d\theta$$

The mean effective magnetic moment,

$$M = \frac{M}{N} = \frac{2\pi \, C \, m \int_0^\pi e^{x\cos\theta} \sin\theta \cos\theta \, d\theta}{2\pi \, C \int_0^\pi e^{x\cos\theta} \sin\theta \, d\theta}$$

$$= m \frac{\int_0^\pi e^{x\cos\theta} \sin\theta \cos\theta \, d\theta}{\int_0^\pi e^{x\cos\theta} \sin\theta \, d\theta}$$

To evaluate the integral, put $\cos\theta = a$ and proceeding exactly as in the case of Langevin theory of dielectrics, we get

$$\frac{M}{m} = \frac{e^x + e^{-x}}{e^x - e^{-x}} - \frac{1}{x}$$

$$= \text{Coth } x - \frac{1}{x} = L(x) \qquad \text{...(4)}$$

where L(x) is called “Langevin function”. When x is small,

$$L(x) = \frac{x}{3} - \frac{x^3}{4 \times 5} + \ldots$$

Under normal experimental conditions x is very small,

$$\therefore \quad L(x) = \frac{x}{3}$$

$$\frac{\overline{M}}{m} = L(x) = \frac{mB}{2KT}, \text{ since } x = \frac{mB}{KT}$$

or $$\overline{M} = \frac{m^2B}{3KT} \quad ...(5)$$

For paramagnetic substances,

$$B = \mu H$$

Hence, $$\overline{M} = \frac{m^2\mu H}{3KT} \quad ...(6)$$

The volume susceptibility of the paramagnetic substance,

$$\chi_{mv} = \frac{m^2\mu n}{3KT}, \left(\sin ce \frac{M}{H} = \chi_m\right), \quad ...(7)$$

where n are the number of molecules per unit volume.

Mass susceptibility is the susceptibility per unit mass of the substance,

$$\chi_m = \frac{n_m \mu\, m^2}{3KT} \quad ...(8)$$

where N_m are the number of molecules per unit mass of the gas.

The molecular susceptibility,

$$\chi_m = \frac{N_m \mu\, m^2}{3KT} \quad ...(9)$$

where N_m are the number of molecules in a gram molecule (Avogadro's number),

or $$\chi_m = \frac{C_m}{T} \quad ...(10)$$

where C_m is called the "Curie constant" given by,

$$C_m = \frac{N_m{}^2 \mu\, m^2}{3R}, \text{ since } K = \frac{R}{N_m} \quad ...(11)$$

χ^m is referred as "molar susceptibility" *i.e.,* susceptibility for one gram molecules of the substance.

Thus, we see that susceptibility depends on the temperature. However, the total magnetisation of a paramagnetic material must include the induced moment, and adding their contribution given in Eq. (7) to Eq. (5), we get,

$$\vec{M} = \left(\frac{m^2}{3KT} - \frac{e^2}{6m_e} \Sigma r^2 \right) \vec{B} \quad \text{...(11a)}$$

Similarly, the total susceptibility of a paramagnetic material must include the induced susceptibility.

Failure of Langevin Theory

Langevin theory could not explain a more complicated dependence of susceptibility upon temperature exhibited by several paramagnetics such as highly compressed and cooled gases, very concentrated solutions of salt, solid salts and crystals. Langevin theory also did not explain the intimate relation observed between para and ferro-magnetism.

WEISS MOLECULAR FIELD THEORY OF PARAMAGNETISM

As stated above, the Landing theory did not explain the intimate relation observed between para and ferro-magentism. In an attempt to explain these points, Weiss in 1907, improved upon Langevin's treatment by the introduction of a new concept of internal molecular field. Weiss assumed that, since in a real gas the molecules are mutually influenced by their magnetic moments, there should exist within the gas a molecular field, produced at any point by all the molecules in the neighbourhood, proportional to and acting in the same sense as the magnetisation vector $\vec{M}$. Therefore, we have

$$\text{Molecular field } \mu \text{ Magnetisation } \vec{M}$$

$$= \gamma \vec{M}$$

where γ is a constant (molecular field coefficient). Then the effective field may be regarded as the vector sum of the external applied field $\vec{B}$ and internal molecular field *i.e.,*

$$B_{eff} = \vec{B} + \gamma\vec{M} \quad \text{...(1)}$$

Replacing B in equation (1) by B_{eff}, we get,

$$M = \frac{m^2 (B + \gamma M)}{3KT}$$

The magnetic moment in a gram molecule,

$$M = \frac{N_m m^2 (B + \gamma M)}{3KT}$$

where N_m are the number of molecules in a gram molecule. Rearranging the terms

$$M\left(1-\frac{N_m m^2\gamma}{3KT}\right)=\frac{N_m m^2 B}{3KT}$$

$$M=\frac{N_m m^2 B}{3KT\left(1-\frac{N_m m^2\gamma}{3KT}\right)}$$

$$=\frac{N_m m^2 B}{(3KT-N_m m^2\gamma)}$$

$$=\frac{N_m m^2 B}{3K\left(T-\frac{N_m m^2\gamma}{3K}\right)}$$

But $B=\mu H$

$$\text{Therefore, } M=\frac{N_m m^2\mu H}{3K\left(T-\frac{N_m m^2\gamma}{3K}\right)}$$

Molecular susceptibility,

$$\chi_m=\frac{M}{H}=\frac{N_m m^2\gamma}{3K\left(T-\frac{N_m m^2\gamma}{3K}\right)}$$

$$=\frac{N_m^2 m^2\mu}{3R\left(T-\frac{N_m^2 m^2\gamma}{3R}\right)},\ \text{since } K=\frac{R}{N_m}$$

$$C_m=\frac{N_m^2 m^2\mu}{3R} \text{ is Curie constant}$$

$$\text{and } \pi=\frac{N_m^2 m^2\gamma}{3R}$$

$$\text{then, } \chi_m=\frac{C_m}{T-\theta} \qquad ...(2)$$

This is the Curie Weiss law. According to this law, the susceptibility of a paramagnetic substance having molecular field varies inversely as the excess of temperature not from absolute zero, as in the case of Curie's law but from a certain critical temperature, the Curie point. The Curie point is able to explain the intimate relation observed between paramagnetic and ferromagnetic substances. Below the Curie point q, the ferromagnetism exists, but above the Curie point, the ferromagnetic properties disappear and the substance becomes paramagnetic.

DOMAIN THEORY OF FERROMAGNETISM

Magnetic effects in most substances are weak. However, a group of substances know as ferromagnetic material exhibit strong magnetic effects. The ferromagnetism is really an extreme case of paramagnetism. If the permanent dipoles, generally those resulting from electron spin, are very close together in the substance, there proves to be an effect (explained only by Quantum Theory), called "exchange", which results in a strong tendency for the spins of adjacent atoms or molecules to line up parallel to each other, even in the absence of an external field.

Such a parallel orientation can extend, in an unmagnetised substance, over volumes of a considerable scale on an atomic order of magnitude (with physical dimensions of the order of 10^{-3} to 10^{-4} cm), though a small volume by ordinary standards. Such a volume is called a "domain", and an ordinary unmagnetised ferromagnetic substance consists of many domains. A strong permanent moment is associated with each domain. The domains are oriented in different directions. In the presence of an external magnetic field, the domains change the orientation of their permanent moments, lining them up with the external fields. As the applied field becomes stronger, the moment increases, until finally with a very large external field the moment reaches a limit when all moments are parallel. This limit is called the "saturation moment".

Reversing the field reverses the moments but there is an effect similar to friction, hindering this reorientation, so that by the time the external field is reduced to zero, the substance still retains a considerable moment. This is the origin of "permanent magnetism". It the external field is reversed alternately between the one direction and other, the moment lags behind the field, resulting in the phenomenon of the "hysteresis".

These properties of ferromagnetics substances are very complicated in contrast to the diamagnetic and paramagnetic substances, in which the moment is proportional to the field. The ferromangetism decreases as the temperature increases because the individual domains lose their moments at a critical temperature called the Curie temperature (about 770°C for iron).

Above the Curie temperature the substance becomes paramagnetic, but the moment instead of being proportional to 1/T where T is the absolute temperature, is proportional to 1/T – θ, where θ is the Curie temperature, is proportional of this decrease of ferromagnetism as the temperature increases is that the tendency towards orientation of the magnetic moments which lines them up at low temperature is opposed b thermal agitation at high temperature.

Thus, the domain theory of ferromagnetism furnishes a clear picture of the phenomena that occur in a ferromagnetic material.

Table 1 : Three Magnetic Vectors

Name	Symbol	Associated with	Boundary condition
Magnetic induction or flux density	$\vec{B}$	All currents	Normal component continuous
Magnetic field strength	$\vec{H}$	True currents only	Tangential component continuous
Magnetisation (Magnetic dipole moment per unit volume)	$\vec{M}$	Magnetisation currents only	Vanishes in a vacuum

Defining equations for $\vec{B}\,\vec{F} = q\vec{v}\times\vec{B} = I\vec{l}\times\vec{B}$

General relation among the there vectors $\vec{B} = \mu_o\vec{H} + \mu_o\vec{M}$

Ampere's law when magnetic materials are present $\oint\vec{H}.d\vec{l} = I$

(I = true currents only)

Empirical relations for certain magnetic materials $\vec{B} = \mu_r\mu_o\vec{H}$

$\vec{M} = (\mu_r - 1)\vec{H}$

COMPARISON OF STATIC ELECTRIC AND MAGNETIC FIELDS

In order to understand clearly, it is necessary to compare electric and magnetic fields and to note both their differences and their similarities.

1. Analogy of $\vec{B}$ *to* $\vec{E}$

By comparison, we see certain analogies. For example we, have noted in electric fields that $\vec{E}$ is defined in terms of force, while in magnetic fields $\vec{B}$ is also defined in terms of force. Hence, $\vec{B}$ may be considered as the magnetic quantity that is analogous to the electric field intensity $\vec{E}$.

2. Analagy of $\vec{H}$ *to* $\vec{D}$

For example in a capacitor $\vec{D}$ is directly related to the electric charge on the plates and is independent of the medium, while near a long current-carrying wire $\vec{H}$ (By $\vec{B} = \mu\vec{H}$) is directly related to the current $\left(B = \dfrac{\mu_0 I}{2\mu R}\right.$ in vacuum and in any medium of permeability μ, $B = \dfrac{\mu I}{2\pi R}$, hence $\left. H = \dfrac{I}{2\pi R}\right)$ and is independent of the medium. Thus, $\vec{D}$ and $\vec{H}$ may be regarded as analogous quantities.

3. Analogy of $\vec{H}$ *to* $\vec{E}$

However, in some other instances we may note an analogy of $\vec{H}$ to $\vec{E}$. For example, the line integral around a closed path of the total electric field intensity $\vec{E}$ is emf ($\oint \vec{E}.d\vec{l} = \xi$), while the line integral of $\vec{H}$ around a closed path is magnetomotive force (mmf $= \oint \vec{H}.d\vec{l}$). Thus $\vec{E}$ and $\vec{H}$ may be regarded as analogous quantities.

4. Analogy of $\vec{B}$ *to* $\vec{D}$

Furthermore in some instances we may note an analogy of $\vec{B}$ to $\vec{D}$. For example, the divergence relations,

$$\nabla.\vec{D} = r \text{ and } \nabla.\vec{B} = 0, \text{ everywhere.}$$

For charge free space (r = 0)

$$\nabla.\vec{D} = 0, \; \nabla.\vec{B} = 0$$

We note a mathematical similarity of $\vec{B}$ to $\vec{D}$.

Important Comment

If the analogy of $\vec{H}$ to $\vec{E}$ and $\vec{B}$ to $\vec{D}$ is pursued, it is possible to achieve a formal or mathematical symmetry between many of the electric and magnetic field equations. *However, electric and magnetic fields are* *fundamentally different, and the analogy of $\vec{B}$ to $\vec{E}$ and $\vec{H}$ to $\vec{D}$ has more meaningful physical significance. Static electric fields are due to electric charge, a scalar quantity while static magnetic fields are due to electric current, a vector quantity.* The electric and magnetic field relations developed are summarised.

Table 2 : A comparison of Static Electric and Magnetic Field Equations

Description of equation	Electric Fields	Magnetic Fields
Force	$\vec{F} = q\vec{E}$	$d\vec{F} = I d\vec{l} \times \vec{B}$ $\vec{F} = Q_m \vec{B}$
Basic relations for lamellar or solenoid field	$\nabla \times \vec{E}_c = 0$	$\nabla . \vec{B} = 0$
Derivation from scalar or vector potential	$E_c = -\nabla V$ $V = \frac{1}{4\pi\varepsilon_0} \int \frac{\rho}{r} dv'$	$\vec{B} = \nabla \times \vec{A}$ $\vec{A} = \frac{\mu_0}{4\pi} \int_v \frac{\vec{j}}{r} dv'$
Relation for $\vec{D}$ and $\vec{H}$	$\vec{D} = \in \vec{E}$ $\nabla . \vec{D} = r_f$	$\vec{H} = \frac{\vec{B}}{\mu}$ $\nabla \times \vec{H} = \vec{j}$
Energy density	$W_e = \frac{1}{2} \in E_2$	$W_m = \frac{1}{2} \frac{B^2}{\mu}$
Capacitance and inductance	$C = \frac{\text{Charge}}{\text{Potential diff.}}$	$L = \frac{\text{Flux linkage}}{\text{Current}}$
Closed path of integration	$\oint \vec{E} . d\vec{l} = \text{emf}$ $\oint \vec{E} . d\vec{l} = 0$ and	$\oint \vec{H} . d\vec{l} = NI = \text{mmf}$ $\oint \vec{H} . d\vec{l} = 0$ (no current enclose)
Derivation from scalar potentials	$\vec{E}_c = -\nabla V$	$H = -\nabla V_m$ (In current free region)

*$\vec{E}_c$ is the static electric field intensity (due to charges).

$\vec{E}$ (without subscript) implies that emf producing fields (not due to charges) may also be present.

The analogy of $\vec{B}$ to $\vec{E}$ and $\vec{H}$ to $\vec{D}$ will be noted in many of the equations while the other analogy may be observed in a few of the relations. The relations apply to static and slowly time varying situations.

They also apply in more rapidly time varying situations with the exception of those relations involving the curl or the line integral ($\vec{H} = -\nabla V_m$ also does not apply in rapidly time varying case).

COMPARISON OF ELECTRIC AND MAGNETIC RELATIONS INVOLVING POLARISATION AND MAGNETISATION

It is interesting to compare the magnetic relations where magnetisation $\vec{M}$ is present with the corresponding electric relations where polarisation $\vec{P}$ is present. This is done in Table 3.

Table 3: Comparison of Equations involving polarisation $\vec{P}$ and Magnetisation $\vec{M}$

Description of Equation	Electric case	Magnetic case
Dipole moment relation	$\vec{P} = \frac{\vec{P}}{v} = \frac{Q\vec{d}}{v}$	$\vec{M} = \frac{\vec{m}}{v} = \frac{Q_m\vec{l}}{v}$
Flux density	$D = \left(\varepsilon_o + \frac{\vec{P}}{\vec{E}}\right)\vec{E}$	$\vec{B} = \mu_o\left(1+\frac{\vec{M}}{\vec{H}}\right)\vec{H}$
Permittivity and permeability	$\varepsilon = \varepsilon_o + \frac{\vec{P}}{\vec{E}}$	$\mu = \mu_o\left(1+\frac{\vec{M}}{\vec{H}}\right)$
Relation to polarisation charge density and to equivalent current dainty	$\nabla.\vec{P} = -\rho_p$	$\nabla \times \vec{M} = \vec{J}_m$
Poisson's eqn.	$\nabla 2V = -\frac{\rho}{\varepsilon_o}$ in vacuum $\nabla 2V = -\frac{\rho}{\varepsilon}$ in medium	$\nabla 2H_m = \nabla.\vec{M} = -\nabla.\vec{H}$
Scalar and vector potentials	$V = \frac{1}{4\pi\varepsilon_o}\int_v \frac{\rho - \nabla' P}{r} dv'$	$\vec{A} = \frac{\mu_o}{4\pi}\int_v \frac{\vec{J} + \nabla\times\vec{M}}{r} dv'$
Magnetisation curve	$\vec{P} - \vec{E}$ curve	$\vec{M} - \vec{H}$ curve identical to $\vec{P} - \vec{E}$ curve
Susceptibility	$\epsilon = \epsilon_o (1 + \chi_{er})$ χ_{er} dimensionless	$\mu = \mu_o (1 + z_m)$ z_m is dimensionless

DETERMINATION OF MAGNETIC SUSCEPTIBILITY OF DIA AND PARAMAGNETIC SUBSTANCES

Let a small piece of a given material be suspended in the magnetic field of an electromagnet as shown in Fig. 2.15. Further, let the magnetic field between pole pieces be non-uniform *i.e.,* it varies from place to place. The suspended substance will experience a force. As the diamagnetism is an induced effect *i.e.,* magnetisation is in the opposite direction to the magnetising field, hence if the sample under consideration is of *diamagnetic material it will tend to move to the region of smaller magnetic field.* On the other hand, since for paramagnetic substances the magnetisation is in the same direction as the magnetising field hence if the sample is of *paramagnetic material it will move towards the region of higher magnetic field.* Suppose a force F acts on the sample placed in the non-uniform magnetic work done by this force will be F.dx. *This work done will be equal to the change in potential energy caused by the displacement of the sample.* Let a small volume v of the given material be situated at the place where B is magnetic field and the permeability of the medium (air) occupying the field when the sample is outside the field is μα. Tηε energy density of the magnetic field on the medium (air) is given by eq. (1), *i.e.,*

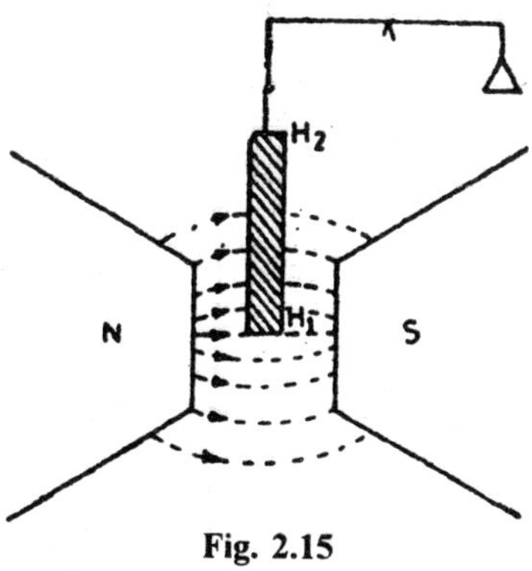

Fig. 2.15

$$\frac{B^2}{2\mu_a} = \frac{\mu_a H^2}{2}$$

Thus, the volume v which is to be occupied by the sample contains magnetic energy of amount,

$$U_1 = \frac{\mu_2 H^2}{2} v$$

Now, the sample of permeability m is placed in the magnetic field. Then the magnetic energy contained by the volume v is

$$U_2 = \frac{\mu H^2}{2} v.$$

Therefore, when a volume v of the given material replaces the same volume of air, it contains an *excess amount of energy given by,*

$$U_2 - U_1 = \frac{(\mu - \mu_a) v H^2}{2}$$

As the magnetising field H is not uniform, then the sample experiences a force and consequently moves through a distance dx. The change in the magnetic energy during this motion is,

$$\frac{d}{dx}\left[\frac{(\mu-\mu_a)vH^2}{2}\right]dx = \frac{(\mu-\mu_a)}{2} v2H \frac{dH}{ddx} dx$$

$$= (\mu - \mu_a)\, vH \frac{dH}{dx} dx$$

$$\frac{\mu}{\mu_o} = \chi_m + 1 \text{ or } \mu = \mu_o (\chi_m + 1)$$

where χ_m is magnetic susceptibility of material. For air we have

$$\mu_a = \mu_o (\chi_{ma} + 1)$$

where χ_{ma} is the magnetic susceptibility of air. Therefore,

$$(\mu - \mu_a) = \mu_o [\chi_m - \chi_{ma}]. \text{ Hence}$$

$$(\mu - \mu_a)\, vH \frac{dH}{dx} dx = \mu_o (\chi_m - \chi_{ma})\, vH \frac{dH}{dx} dx$$

This change in magnetic energy is equal to the work done on the sample. Thus,

$$F dx = \mu_o [\chi_m - \chi_{ma}]\, vH \frac{dH}{dx} dx$$

or $$F = \chi_o [\chi_m - \chi_{ma}]\, vH \frac{dH}{dx} \qquad ...(1)$$

By using this equation we can determine the susceptibility χ_m of a dia or paramagnetic material, provided magnetic susceptibility of air is known. We shall discuss two methods.

1. Gouy Method when the substance is available in the form of solids.

2. Quincke's method when the substance is available in the form of liquids.

1. Gouy Method for Measuring Susceptibility of Solids

The sample under test is taken in the form of thin rod and is suspended from one arm of a sensitive micro balance in the magnetic field of electromagnet. The field will vary rapidly along the vertical direction, hence the pole pieces are wedge shaped. If the pole pieces are sufficiently wide apart, the variation in the field in the horizontal direction will be small. Hence the field is uniform along horizontal direction.

As the vertical dimension of the pole pieces is small, the magnetic field falls rapidly in the vertical direction. Procedure is as follows.

Put the specimen at the central part of the field where the field is uniform horizontally. First the specimen is balanced by putting weights on the other arm of the balance. Now the magnetic field is switched on and the force (upward or downward depending on whether the specimen is dia or paramagnetic) acts on the specimen due to the non-uniform field in vertical direction. This force is counter balanced by adjusting the weights to balance the specimen. If the weights added or removed in the second case be m, then the force experienced by specimen due to magnetic field is mg. If A is the area of the cross-section of rod, then volume of dx length of the rod is Adx where x is measured vertically from upper end. Then the force on dx length of the rod is by eq. (1),

$$dF = \mu, [\chi_m - \chi_{ma}] \, AdxH \frac{dH}{dx}, \; [v = Adx]$$

Therefore, the force experienced by the whole rod is

$$F = \mu_o [\chi_m - \chi_{ma}] \, A \int_{H2}^{H1} HdH$$

$$F = \frac{1}{2} \mu_o (\chi_m - \chi_{ma}) \, A \, (H_1^2 - H_2^2) \qquad ...(2)$$

where H_1 and H_2 are the values of magnetic field at the lower and upper ends of the rod.

But, F = mg, hence

$$mg = \frac{1}{2} \mu_o \, A \, (\chi_m - \chi_{ma}) \, (H_1^2 - H_2^2)$$

As magnetic field falls very rapidly upward from the centre, hence H_2 is very small it can be neglected. Therefore,

$$mg \approx \frac{1}{2} \mu_o A (\chi_m - \chi_{ma}) H_1^2 \qquad ...(3)$$

If χ_{ma} is neglected as it is very small is comparison to cma then,

$$mg = \frac{1}{2} \mu_o A \chi_m H_1^2 \qquad ...(4)$$

Hence by using eq. (3) or (4) χ_m can be calculated. Magnetic field H_1 is measured by fluxmeter.

2. Quincke's Method for Measuring Susceptibility of Liquids

Quincke directly used Gouy method for liquids. The liquid under test is taken in a U tube, one end of which is capillary section and the other end as wide tube.

As shown in Fig. 2.16, the capillary section is placed between the pole pieces of electromagnet. The liquid meniscus is placed in the centre of the pole pieces where the field is uniform. When the current energising the electromagnet is switched on under the action of magnetic field the level of meniscus falls or rises depending on whether liquid is diamagnetic or paramagnetic. This change in the height of the meniscus is measured by a travelling microscope. If it is h, then h is the height of the liquid column supported by the force due to non-uniform magnetic field.

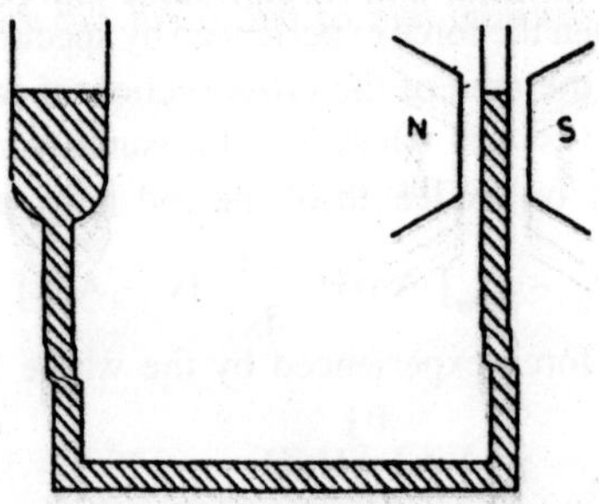

Fig. 2.16

If ρ be the density of the liquid and σ be that of air , then change in pressure of the liquid corresponding to the change h is $(\rho - \sigma)$ gh. Therefore, force acting downwards on the liquid

$$= (\rho - \sigma) gh A \qquad ...(5)$$

where A is area of cross-section of the capillary tube.

The vertical force exerted by magnetic field is given by eqn. (1) is,

$$dF = \mu_o (\chi_m - \chi_{ma}) Adx\, H \frac{dH}{dx}$$ where v has been replaced by

Adx. The total force can be obtained by integrating the above equation from H = 0 *i.e.*, the initial zero magnetic field to $H = H_m$, the final value of magnetic field,

$$F = \mu_o [\chi_m - \chi_{ma}] \int_{H=0}^{H=H_m} Adx \frac{HdH}{dx}$$

or $$F = \frac{1}{2} \mu_o (\chi_m - \chi_{ma}) A H_m^2 \quad ...(6)$$

Equating (5) and (6), we get,

$$(\rho - \sigma) gh = \frac{1}{2} \mu_o (\chi_m - \chi_{ma}) H_m^2 \quad ...(7)$$

If the susceptibility of air χ_{ma} is known we can calculate χ_m from equation (7). The H_m can be measured by fluxmeter.

SOLVED EXAMPLES

Example 1:

If, H = 1000 amp./metre then calculate the magnetic potential difference between the ends of the bar and total flux.

Solution:

$V_m = \int \vec{H}.d\vec{l}$, as H is along the length of the bar, then $\vec{H}.d\vec{l}$ = Hdl. Magnetic potential between the ends,

$$V_m = \int_0^{0.1} Hdl = 1{,}000 \times 0.1 = 1000 \text{ amp.}$$

The flux density $B = \mu_1 H = 500\mu_o \times 1000$ weber/metre^2

$= 6.28 \times 10^{-1}$ weber/metre^2

The total flux = BA

$= 6.28 \times 10^{-1} \times 15 \times 10^{-4}$

$= 9.4 \times 10^{-4}$ weber

Example 2:

In example 2.3 if B = 1 weber/metre^2, then calculate the potential difference between the ends of blocks.

Solution:

From boundary condition the normal component of B is continuous in block 1

$$H_1 = \frac{B}{\mu_1} = \frac{1}{500\mu_o} = 1.59 \times 10^3 \text{ amp.metre}$$

and potential difference

$$V_{m1} = H_1 l_1 = 1.59 \times 10^3 \times 0.1 = 159 \text{ amp.}$$

In block 2,

$$H_2 = \frac{B}{\mu_2} = \frac{1}{20000\mu_o} = 3.97 \times 10^2 \text{ amp/metre}$$

and potential difference V_m between the end faces of block 2,

$$V_{m2} = H_2 l_2 = 3.7 \times 10^2 \times 0.2 = 79.4 \text{ amp.}$$

The total potential difference across both blocks is then given by,

$$V_m = V_{m1} + V_{m2}$$

$$= 159 + 79.4 = 238.4 \text{ amp.}$$

Example 3:

Find the total reluctance and permeance between the ends of parallel-connected rectangular iron $\mu_1 = 500\ \mu_o$ and $\mu_2 = 2000\ \mu_o$

Solution:

Since the blocks are connected in parallel, it is convenient to calculate the total permeance first.

$$\text{Permeance of block 1} = \frac{\mu_1 A_1}{l}$$

$$= \frac{500 \times 4\pi \times 10^{-7} \times 20 \times 10^{-4}}{0.2}$$

$$= 6.28 \times 10^{-6} \text{ henry}$$

$$\text{Permeance of block 2} = \frac{\mu_2 A_2}{l}$$

$$= \frac{2000 \times 4\pi \times 10^{-7} 10 \times 10^{-4}}{0.2}$$

$$= 12.6 \times 10^{-6} \text{ henry}$$

$$\text{Total permeance} = (6.28 + 12.6)\ 10^{-6} \text{ henry}$$

$$= 1.89 \times 10^{-5} \text{ henry}$$

The total reluctance is given by

$$= \frac{1}{1.89 \times 10^{-5}} = 5.3 \times 10^4 \frac{1}{\text{henry}}$$

Example 4:

In the above example 2.6 if H = 1000 amp/metre, calculate the total flux.

Solution:

Since from boundary condition the tangential component of H is continuous that H is same in both blocks, the,

$$B_1 = \mu_1 H = 500\ \mu_o \times 1000 = 0.628 \text{ weber/metre}^2$$

$$\phi B_1 = B_1 A_1 = 0.628 \times 20 \times 10^{-4} = 1.26 \times 10^{-3} \text{ weber}$$

and $B_2 = \mu_2 H = 2000\ \mu_o \times 1000 = 2.52 \text{ weber/metre}^2$

$$\phi B_2 = B_2 A_2 = 2.52 \times 10^{-3} \text{ weber}$$

Total flux $= \phi B_1 + \phi B_2 = 3.78 \times 10^{-3}$ weber

Example 5:

An iron ring have a cross-sectional area 10 cm² and air gap of width d = 2 mm and mean length l = 2πR = 60 cm including the air gap. Find the number of turns required to produce flux density, B = 1 weber/metre², μ_{iron} = 10^{-3} henry/metre.

Solution:

$$\text{Total flux} = \frac{\text{mmt}}{\text{Re luc tan ce}}$$

$$\phi_B = \frac{NI}{\dfrac{d}{\mu_0 A} + \dfrac{l-d}{\mu A}}$$

$$\text{Flux density, } \phi_B = BA = \frac{NI}{\dfrac{d}{\mu_0 A} + \dfrac{l-d}{\mu A}}$$

or $$NI = BA\left(\frac{d}{\mu_0 A} + \frac{l-d}{\mu A}\right)$$

$$= B\left(\frac{d}{\mu_o A} + \frac{l-d}{\mu}\right)$$

$$= 1 \times \left(\frac{2\times10^{-6}}{4\pi\times10^{-7}} + \frac{598\times10^{-6}}{10^{-3}}\right)$$

= 2188 amp. turns.

Example 6:

An iron rod of length 50 cm and cross-sectional area 4 sq. cm. is in the form a closed circular ring. If the permeability of iron is 65 × 10^{-4} h/m. Calculate the number of ampere turns required to produce a flux of 4 × 10^{-4} weber.

Solution:

$$\text{Reluctance} = \frac{l}{\mu A} = \frac{0.5}{65\times10^{-4}\times4\times10^{-4}}$$

$$= 1.92 \times 10^5 \text{ amp. turns/weber}$$

$$\text{mmf} = NI$$

$$\text{Flux} \quad \phi_B = \frac{\text{mmf}}{\text{Reluctance}}$$

$$4 \times 10^{-4} = \frac{NI}{1.92\times10^5}$$

$$NI = 77 \text{ amp turns}$$

Example 7:

The magnetic susceptibility of medium is 948 × 10^{-11}. Calculate the permeability (or absolute permeability) and relative permeability.

Solution:

Hint: $\xi_m = 948 \times 10^{-11}$

Absolute permeability

$$\mu = \mu_o (1 + \xi_m)$$

$$= 4\pi \times 10^{-7} (1 + 948 \times 10^{-11}) \text{ henry/m}$$

$$= 4\pi \times 10^{-7} \times 949 \times 10^{-11} \text{ henry/m}$$

$$= 1.186 \times 10^{-14} \text{ henry/metre}$$

and relative permeability,

$$\mu_r = \frac{\mu}{\mu_0}$$

$$= \frac{4\pi \times 10^{-7} \times 949 \times 10^{-11}}{4\pi \times 10^{-7}}$$

$$= 949 \times 10^{-11}.$$

Example 8:

In the row land ring 2.0 amp current (i_0) is passing in the windings and the number of turns per unit length (n) in the toroid are 10 turns/cm. B measured is 1.00 weber/metre². Calculate (a), $\vec{H}$, (b) $\vec{M}$ and (c) the magnetising current, both when the core is present and when it is removed. (d) What is μ_r^2.

Solution:

(a) As $\vec{H}$ is independent of the core material and is given by Ampere's law

$$\oint \vec{H}.d\vec{l} = I$$

$\vec{H}$ is tangential to l

$$H \oint dl = NI$$

where N are the total number of turns on the toroidal windings and NI is enclosed current.

$$H_l = NI$$

$$H = \frac{N}{l} I = NI$$

$$= (10 \text{ turns/cm}) (2.0 \text{ amp})$$

$$= 2.00 \text{ amp} \times 10 \times 100 \text{ turns/metre}$$

$$= 2000 \text{ amp/metre}$$

(b) The magnitude of $\vec{M}$ is obtained by relation

$$\vec{B} = \mu_0 \vec{H} + \mu_0 \vec{M}$$

$$M = \frac{B - \mu_0 H}{\mu_0}$$

$$= \frac{1-4\pi\times10^{-7}\times2000}{4\pi\times10^{-7}}$$

(c) The effective magnetising current from Eq. (2.19a),

$$i_m = M.\left(\frac{2\pi r}{N}\right) = M/n$$

$$= \frac{7.9\times10^5 \text{ amp / metre}}{2.0\times10^3 \text{ turns / metre}}$$

$$= 390 \text{ amp.}$$

Thus, an additional current of this amount in the windings would produce the same value of B in the absence of a core as that obtained with the core in place.

(d) The permeability

$$\mu_r = \frac{B}{\mu_o H}$$

$$= \frac{1 \text{ weber / metre}^2}{(4\pi\times10^{-7} \text{ web / amp. metre})(2\times10^2 \text{ a mp / met})}$$

$$= 397$$

EXERCISES

1. Distinguish dia, para-and ferro-magnetic materials. Give the Langvin theory of diamagnetism.
2. What are atomic currents? How will you distinguish between magnetic and non-magnetic materials?
3. Prove the magnetic due to orbital motion of an electron must be an integral multiple of eh/4πm.
4. What do you mean by magnetisation? What is difference between uniform and non-uniform magnetisation? Show that for non-uniform magnetisation

   $$\nabla \times \vec{M} = \vec{J}$$

5. Derive the relation between the resultant angular momentum and the resultant magnetic moment of atom. What is gyromagnetic ratio?

6. What are magnetisation curves? Explain residual magnetism, coercive force and hysteresis.
7. Describe the method (B.G.) of tracing of the hysteresis curve for a sample of iron in the form of an anchor ring.
8. Explain the magnetic circuit, magnetomotive force and reluctance. How a magnetic circuit differs from an electric circuit?
9. Distinguish between para. and ferromagnetic materials. Discuss the Langevin theory para-magnetism.
10. What is Curic Weises law?
11. Give domain theory of ferro-magnetism.
12. An iron ring of 10 cm radius and 5 sq. cm. cross-section has an air gap of 2 mm width. A coil is would uniformly over the ring. A flux of 10^{-3} weber is produced in the gap. If a current of 1.82 amp. passes through the coil, find the total number of turns in the coil if the permeability of the core is 65×10^{-4} henry/metre.
13. An iron ring of radius 10 cm and area of cross-section 5 sq. cm. has an air gap 1 cm wide. The ring is would uniformly with coil of 900 turns. A flux of 10^{-4} weber is required in the air gap. Find the strength of current required to produce. Given miron = 65 $\times$ 10^{-4} henry/metre.

3

Electromagnetic Waves and Equations

ELECTROMAGNETIC WAVES IN ISOTROPIC NON-CONDUCTING MEDIA (*i.e.*, DIELECTRICS)

Maxwell's Equation

Let us first write the basic laws with medium of permittivity ε and permeability μ.

Gauss's law for electrostatics is given by,

$$\nabla . \vec{D} = \rho$$

Gauss's law for magnetic field

$$\nabla . \vec{B} = 0$$

Faraday's law, $\nabla \times \vec{E} = -\dfrac{\delta \vec{B}}{\partial t}$

Ampere's law in presence of medium is,

$$\oint \vec{H} . d\vec{l} = I$$

by Stokes's theorem is

$$\nabla \times \vec{H} = \vec{J}$$

But $\quad \vec{B} = \mu \vec{H}$

With this linear relation for $\vec{B}$ and $\vec{H}$, we can write Ampere's law in the presence of medium as,

$$\oint \frac{\vec{B}}{\mu} \cdot d\vec{l} = I$$

or

$$\oint \vec{B}.d\vec{l} = \mu I$$

Then by Stokes' theorem,

$$\nabla \times \vec{B} = \mu \vec{J}$$

Now with the introduction of displacement current $\frac{\delta \vec{D}}{\delta t}$ the Ampere's law is expressed by replacing J by $\vec{J} + \frac{\delta \vec{D}}{\delta t}$. Therefore, we can write,

$$\nabla \times \vec{B} = \mu \left(\vec{J} + \frac{\delta \vec{D}}{\delta t} \right)$$

Thus, the complete equations of electromagnetic theory are as follow:

$$\nabla . \dot{D} = 0 \quad \text{(a)}$$

$$\nabla . \vec{B} = 0 \quad \text{(b)}$$

$$\nabla \times \vec{E} = -\frac{\delta \vec{B}}{\delta t} \quad \text{(c)} \quad ...(1)$$

$$\nabla \times \vec{B} = \mu \left(\vec{J} + \frac{\delta \vec{D}}{\delta t} \right) \quad \text{(d)}$$

with $\vec{D} = \varepsilon \vec{E}$

and $\vec{B} = \mu \vec{H}$

As dielectrics are non-conducting medium, therefore, current $\vec{J} = 0$ and in a homogeneous isotropic medium there is no volume distribution of charge, thus, the charge density $\rho = 0$. Hence the constitutive equations are

$$\vec{D} \text{ keo} \vec{E} = \varepsilon \vec{E}$$

$$\vec{B} = \text{momr} \vec{H} = \mu \vec{H}$$

$$\vec{J} = 0$$

$$\rho = 0$$

Hence the Maxwell's equations for dielectrics medium from equation (1).

$$\nabla.\vec{E} = 0 \qquad (a)$$

$$\nabla.\vec{B} = 0 \qquad (b)$$

$$\nabla \times \vec{E} = -\frac{\delta\vec{B}}{\delta t} \qquad (c) \qquad ...(2)$$

$$\nabla \times \vec{B} = \mu\varepsilon \frac{\delta\vec{E}}{\delta t} \qquad (d)$$

We can obtain the equation of propagation of wave in dielectrics medium by eliminating $\vec{E}$ from Eqns. (2c) and (2d) in the same manner as for free space.

Taking the curl of equation (2d) we have,

$$\nabla \times \nabla \times \vec{B} = \nabla \times \mu_o \frac{\delta\vec{E}}{\delta t}$$

As the medium is isotropic μ and ε remain constant throughout the material of the medium hence can be taken out of the ∇ operation.

Thus,

$$\nabla \times \nabla \times \vec{B} = \mu_o \nabla \times \frac{\delta\vec{E}}{\delta t}$$

$$= \mu\varepsilon \frac{\delta}{\delta t}(\nabla \times \vec{E})$$

Substituting,

$\nabla \times \vec{E}$ from eqn. (2c), we get,

$$\nabla \times \nabla \times \vec{B}\ \mu\varepsilon \frac{\delta}{\delta t}\left(-\frac{\delta\vec{B}}{\delta t}\right)$$

$$= -\mu\varepsilon \frac{\delta^2\vec{B}}{\delta t^2}$$

But, $\nabla \times \nabla \times \vec{B} = \nabla(\nabla.\vec{B}) - \nabla^2\vec{B}$

Since, $\nabla.\vec{B} = 0$

Therefore, we get,

$$\nabla \times \nabla \times \vec{B} = -\nabla^2\vec{B}$$

Substituting in the above equation, we get,

$$-\nabla^2\vec{B} = -\mu\varepsilon \frac{\delta^2 B}{\delta t^2}$$

or $$\nabla^2\vec{B} = \mu\varepsilon \frac{\delta^2\vec{B}}{\delta t^2} \qquad ...(3)$$

Similarly $\vec{B}$ can be eliminated from the equation (2c) and (2d) and we get,

$$\nabla^2\vec{E} = \mu\varepsilon \frac{\delta^2\vec{E}}{\delta t^2} \qquad ...(4)$$

Clearly equations (3) and (4) representing the relation between the space and time variation of magnetic field $\vec{B}$ and electric field $\vec{E}$ are called wave equations for $\vec{B}$ and $\vec{E}$ repetitively in a three dimensional case. If we substitute,

$$\nabla^2 = \frac{\delta^2}{\delta x^2} + \frac{\delta^2}{\delta y^2} + \frac{\delta^2}{\delta z^2}$$

$$\vec{B} = \hat{i}B_x + \hat{j}B_y + \hat{k}B_z$$

Then, $$\left(\frac{\delta^2}{\delta x^2} + \frac{\delta^2}{\delta y^2} + \frac{\delta^2}{\delta z^2}\right)(\hat{i}B_x + \hat{j}B_y + \hat{k}B_z)$$

$$= \mu\varepsilon \frac{\delta^2}{\delta t^2}(\hat{i}B_x + \hat{j}B_y + \hat{k}B_z)$$

Similar expression for $\vec{E}$ is obtained if we analyse eq. (4). Each component of $\vec{B}$ and $\vec{E}$ is identical with the familiar wave equation

$$\frac{\delta^2 y}{\delta x^2} = \frac{1}{v^2}\frac{\delta^2 y}{\delta t^2}$$

Here, therefore $\mu\varepsilon$ has the same significance as the $\frac{1}{v^2}$.

Dimensionally equation (3) is,

$$\frac{1}{\text{metre}^2}\cdot\frac{\text{weber}}{\text{metre}^2} = \mu\varepsilon \frac{1}{\text{sec}^2}\cdot\frac{\text{weber}}{\text{metre}^2}, \text{ since}$$

$$B = \frac{\text{Weber}}{\text{metre}^2} \text{ so that } \frac{1}{\mu \in} = \frac{\text{meter}}{\text{sec.}}$$

Same result is obtained if we analyse equation (4).

Thus, it appears that $1\sqrt{\mu\varepsilon}$ *as the dimensions of velocity and is in fact the velocity with which the wave propagates in the dielectrics,* if represented by v, then

$$v = \frac{1}{\sqrt{\mu\varepsilon}} \qquad ...(5)$$

Thus we can conclude that the field vectors $\vec{B}$ and $\vec{E}$ propagates as waves with velocity $1/\sqrt{\mu\varepsilon}$. The velocity is a characteristic of the medium depending on the constant μ and ε.

INDEX OF REFRACTION

Since $\varepsilon = \varepsilon_o k$

and $\mu = \mu_o \mu_r$

where k is relative permittivity and μ_r is relative permeability

$$\text{Then, } v = \frac{1}{\sqrt{\mu\varepsilon}} = \frac{1}{\sqrt{\mu_o \mu_o \varepsilon_o k}} \qquad ...(1)$$

$$= \frac{1}{\sqrt{\mu_o \varepsilon_o} \cdot \sqrt{\mu_r k}}$$

by equation (1), $1/\sqrt{\mu_o \varepsilon_o} = C$ (velocity of light), hence,

$$v = \frac{C}{\sqrt{\mu_r k}}$$

$$= \frac{C}{\eta} \qquad ...(2)$$

where η is the refractive index of the dielectric which is always greater than unity. *Thus velocity of a wave in a dielectric is less than in free space.* For non-magnetic (or non-ferrous) medium, permeability $\mu_r = 1$ refractive index,

$$h = \frac{C}{v} = \sqrt{\mu_r k} = \sqrt{k} \qquad ...(3)$$

SOLUTIONS OF THE WAVE EQUATIONS

Let us now discuss the solutions of wave equations (1) and (2). There are many solutions corresponding to different kinds of electromagnetic waves such as plane, spherical and cylindrical waves. We shall consider only uniform plane wave. Therefore, we shall assume that $\vec{E}$ and $\vec{B}$ are functions of only one space coordinate along which the wave travels and of the time t and there is no variation in the component of $\vec{E}$ and $\vec{B}$ in any other direction. If $\vec{E}$ and $\vec{B}$ vary along x-axis, then

$$\vec{E} = \vec{E}\,(x, t) \text{ and } \vec{B} = \vec{B}\,(x, t) \qquad ...(1)$$

and all derivatives of the components of $\vec{E}$ and $\vec{B}$ with respect to y and z are zero. Thus,

$$\frac{\delta E_y}{\delta y} = 0,\ \frac{\delta E_z}{\delta z} = 0,\ \frac{\delta B_y}{\delta y} = 0 \text{ and } \frac{\delta B_z}{\delta z} = 0 \qquad ...(2)$$

Then eqs. (1) and (2), *i.e.*,

$$\nabla.\vec{E} = 0 \text{ and } \nabla.\vec{B} = 0$$

written in cartesian coordinates are,

$$\nabla.\vec{E} = \frac{\delta E_x}{\delta x} + \frac{\delta E_y}{\delta y} + \frac{\delta E_z}{\delta z} = 0$$

$$\nabla.\vec{B} = \frac{\delta B_x}{\delta x} + \frac{\delta B_y}{\delta y} + \frac{\delta B_z}{\delta z} = 0 \qquad ...(3)$$

for plane wave by equation (3),

$$\frac{\delta E_x}{\delta x} = 0 \qquad ...(4a)$$

$$\frac{\delta B_x}{\delta x} = 0 \qquad ...(4b)$$

Equations (4a) and (4b) indicate that E_x and B_x cannot be a function of x which is the direction of propagation. This means E_x and B_x are constant in x-direction. However, we are not interested in uniform fields since these are not the part of any wave equation, therefore, we shall set,

$$E_x = 0 \text{ and } B_x = 0 \qquad ...(4c)$$

Hence the vector $\vec{E}$ and $\vec{B}$ can have only y and z components *i.e., it has no "longitudinal" component but has only "transverse" components. Thus,*

$$\vec{E} = \hat{j}E_y + \hat{k}E_z$$

$$\vec{B} = \hat{j}B_y + \hat{k}B_z \qquad ...(4d)$$

To be more clear, the Maxwell's electromagnetic waves are purely transverse and never longitudinal. Also from equations (1), we have,

$$\nabla \times \vec{E} = -\frac{\delta \vec{B}}{\delta t}$$

$$\begin{vmatrix} \hat{i} & \hat{j} & \hat{k} \\ \frac{\delta}{\delta x} & \frac{\delta}{\delta y} & \frac{\delta}{\delta z} \\ E_x & E_y & E_z \end{vmatrix} = -\left(\hat{i}\frac{\delta B_x}{\delta t} + \hat{j}\frac{\delta B_y}{\delta t} + \hat{k}\frac{\delta B_z}{\delta t}\right)$$

On equating the components separately, we get,

$$\frac{\delta E_z}{\delta y} - \frac{\delta E_y}{\delta z} = -\frac{\delta B_x}{\delta t}$$

$$\frac{\delta E_x}{\delta z} - \frac{\delta E_z}{\delta x} = -\frac{\delta B_y}{\delta t}$$

$$\frac{\delta E_y}{\delta z} - \frac{\delta E_x}{\delta y} = -\frac{\delta B_z}{\delta t}$$

Now remembering that $E_x = 0$

and $\frac{\delta}{\delta y} = \frac{\delta}{\delta z} = 0$, we get,

$$\frac{\delta B_x}{\delta t} = 0 \qquad ...(4e)$$

$$\frac{\delta E_z}{\delta x} = \frac{\delta B_y}{\delta t} \qquad ...(4f)$$

$$\frac{\delta E_y}{\delta x} = -\frac{\delta B_z}{\delta t} \qquad ...(4g)$$

Similarly from eq. (3d), *i.e.,*

$$\nabla \times \vec{B} = \mu\varepsilon \frac{\delta \vec{E}}{\delta t}$$

can be written in components, *i.e.,*

$$\frac{\delta B_z}{\delta y} - \frac{\delta B_y}{\delta z} = \mu\varepsilon \frac{\delta E_x}{\delta t}$$

$$\frac{\delta B_x}{\delta z} - \frac{\delta B_z}{\delta x} = \mu\varepsilon \frac{\delta E_y}{\delta t}$$

$$\frac{\delta B_y}{\delta x} - \frac{\delta B_x}{\delta y} = \mu\varepsilon \frac{\delta E_z}{\delta t}$$

Again remembering that $B_x = 0$ and $\frac{\delta}{\delta y} = \frac{\delta}{\delta z} = 0$, we get,

$$\frac{\delta E_x}{\delta t} = 0 \qquad \text{...(4h)}$$

$$\frac{\delta B_z}{\delta x} = -\mu\varepsilon \frac{\delta E_y}{\delta t} \qquad \text{...(4i)}$$

$$\frac{\delta B_y}{\delta x} = \mu\varepsilon \frac{\delta B_z}{\delta t} \qquad \text{...(4j)}$$

The equations [3(f, g, i, j)] clearly represent that E_z is associated with B_y and E_y is associated with B_z. Thus E_z and B_y are quite independent of E_y and B_z. Thus we can write,

$$\frac{\delta E_y}{\delta x} = -\frac{\delta B_z}{\delta t} \qquad \text{...(4k)}$$

and $$\frac{\delta B_z}{\delta x} = -\mu\varepsilon \frac{\delta E_y}{\delta t} \qquad \text{...(4l)}$$

Differentiating equation (4k) with respect to x and eqn. (3) with respect to t, we get

$$\frac{\delta^2 E_y}{\delta x^2} = -\frac{\delta^2 B_z}{\delta x \delta t}$$

and $$\frac{\delta^2 B_z}{\delta x \delta t} = -\mu\varepsilon \frac{\delta^2 E_y}{\delta t^2}$$

hence, $$\frac{\delta^2 E_y}{\delta x^2} = \mu\varepsilon \frac{\delta^2 E_y}{\delta t^2} \qquad \text{...(5a)}$$

or $$\frac{\delta^2 E_y}{\delta t^2} = \frac{1}{\mu\varepsilon} \frac{\delta^2 E_y}{\delta x^2} \qquad \text{...(5b)}$$

The velocity of wave $v = 1/\sqrt{\mu\varepsilon}$

The wave equation (3a) is a linear partial differential equation of the second order disrobing the propagation of the y-component of the electric field $\vec{E}$ along the x-axis with constant velocity v.

For simplifying the calculations let us choose our coordinate axes such that y-axis is parallel to the vector $\vec{E}$. We can do this since E_y and B_z are quite independent of E_z and B_y. Thus we suppose that $E_z = 0$ and $B_y = 0$, then

$$\vec{E} = \hat{j}E_y \text{ and } \vec{B} = \hat{k}B_z$$

Here, as we are interested in electromagnetic waves in which the field is simply periodic function of time *i.e.,* in general the time dependence $\vec{E}$ and $\vec{B}$ is $e^{j\omega t}$ for single fixed frequency.

Thus we write in equation (5a),

$$E_y(x, t) = f(x)\, e^{j\omega t} \qquad ...(6)$$

After substitution in (5a), we get,

$$\frac{d^2 f}{dx^2} = -\frac{\omega^2}{v^2} f \qquad ...(7a)$$

This Equation has two solutions,

$$f(x) = e^{j\omega x/v}$$

and $f(x) = e^{-j\omega x/v}$

Hence a general solution for (3) is

$$E_y = E_{oy}\left[e^{j\omega(t+x/v)} + e^{j\omega(t-x/v)}\right] \qquad ...(7b)$$

Therefore, either term alone is a solution or the sum as in eqn. (7b) is a solution. We can choose 1st term of equation (7b) and take the imaginary part as solution then we get,

$$E_y = E_{oy} \sin \omega \left(t + \frac{x}{v}\right) \qquad ...(8)$$

The significance of eqn. (8) can be illustrated by evaluating E_y as a function of x for several values of time t

At $t = 0$, $E_y = \sin \omega \dfrac{x}{v}$, hence at $x = 0$, $E_y = 0$

At $t = \dfrac{T}{4}$, $\omega t = \dfrac{2\pi}{T}\dfrac{T}{4} = \dfrac{\pi}{2}$ and $E_y = \sin\left(\dfrac{\pi}{2} + \dfrac{\omega x}{v}\right)$

Hence at $x = 0$, $E_y = +1$, The maximum value of E_y is unity.

At $t = \frac{T}{2}$, $\omega t = \frac{2\pi}{T} \cdot \frac{T}{2} = \pi$ and $E_y = \sin\left(\pi + \frac{\omega x}{v}\right)$

Hence at $x = 0$, $E_y = -1$

The curves are shown in Fig. 3.1(a)

Focussing our attention on the crest of one of the waves as indicated by the point P, we note that as time progresses, P moves to the left. Thus we can conclude that equation (8) is representing a wave travelling to the left *i.e.,* negative x-axis. The point P is a point of constant phase and is characterised by the condition $x + vt =$ const., taking the time derivatives $\frac{dx}{dt} + v = 0$

If we take, $E_y = E_{oy} \sin \omega \left(t - \frac{x}{v}\right)$

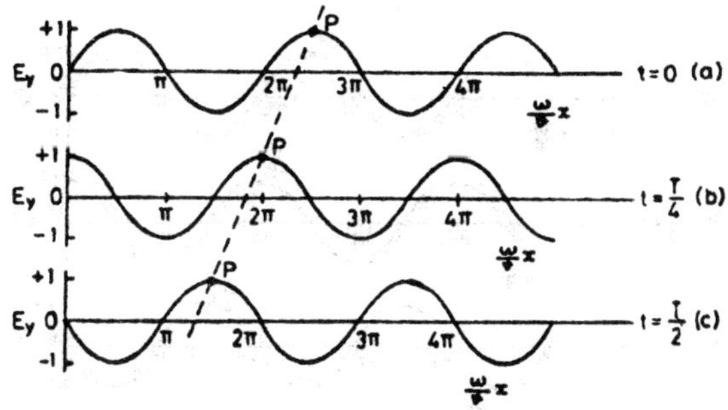

Fig. 3.1

Then proceeding in the same manner we found that it represents the wave travelling to the right positive x-axis. To "summaries" the negative sign in t ±x/y is associated with a wave to the right, while a positive sign is associated with a wave to the left. Other equivalent solution can also be used such as,

$$E_y = E_{oy} \cos \omega(t - x/v) \text{ or } E_y = E_{oy} \cos\left(\frac{\omega}{v}x - \omega t\right)$$

Again for simplifying the calculations, let us consider only the wave travelling in the positive direction of the x-axis, therefore, we have,

$$E_y = E_{oy} \sin \omega(t - x/v) \qquad \text{...(9a)}$$

Now we can calculate B_z by substituting E_y in eq. (6k)

$$\frac{\delta}{\delta x}[E_{oy} \sin \omega(t - x/v)] = -\frac{\delta}{\delta t} B_z$$

$$-E_{oy} \cos \omega\left(t - \frac{x}{v}\right) \cdot \frac{\omega}{v} = -\frac{\delta}{\delta t} B_z$$

Integrating with respect to t, we get,

$$E_{oy} \sin \omega(t - x/v) \cdot \frac{1}{v} = B_z$$

or $$B_z = B_{oz} \sin \omega(t - x/v) \qquad ...(9b)$$

where $B_{oz} = E_{oy}/v$ ...(10)

Hence ratio of eqns. (9a) and (9b)

$$\frac{E_y}{B_z} = \frac{E_{oy}}{B_{oz}} = v$$

or $$\frac{E_y}{B_z} = \frac{1}{\sqrt{\mu\varepsilon}} \qquad ...(11a)$$

If the medium is free space then,

$$\frac{E_y}{B_z} = \frac{1}{\sqrt{\mu_o \varepsilon_o}} \qquad ...(11b)$$

The value of E_y/B_z is real and depends on the medium in which wave is travelling. Equations (9a) and (9b) represent linearly polarised place harmonic wave travelling along the +x axis in which $\vec{E}$ is everywhere parallel to y-axis and $\vec{B}$ is parallel to z-axis. As the ratio E_y/B_y is a real positive constant, hence the two vectors $\vec{E}$ and $\vec{B}$ are in phase when $\vec{E}$ has its maximum value directed along +y direction, $\vec{B}$ has its maximum value directed along +z-axis.

By equation (11a)

$$B_z = \sqrt{\mu \in} E_y$$

Now multiplying the left side by $\hat{k}$ and right side by its equivalent $\hat{i} \times \hat{j}$, we get,

$$\hat{k} B_z = \sqrt{\mu\varepsilon} . \hat{i} \times \hat{j}\ E_y$$

or $$\vec{B} = \sqrt{\mu\varepsilon}\,\hat{i}\times\vec{E} \qquad \text{...(12a)}$$

Similarly, $E_y = \dfrac{1}{\sqrt{\mu\varepsilon}} B_z$

Multiplying left side by $\hat{j}$ and right side its equivalent $\hat{k}\times\hat{i}$, we get,

$$\hat{j}E_y = \frac{1}{\sqrt{\mu\varepsilon}}\,\hat{k}\times\hat{i}\,B_z$$

$$= -\frac{1}{\sqrt{\mu\varepsilon}}\,\hat{i}\times\hat{k}\,B_z$$

$$\vec{E} = -\frac{1}{\sqrt{\mu\varepsilon}}\,\hat{i}\times\vec{B} \qquad \text{(12b)}$$

The vectors $\vec{E}$ and $\vec{B}$ are perpendicular to one another, the directions of $\vec{E}$ and $\vec{B}$ are related to the direction of propagation by the rule of "right hand screw". The vectors $\vec{E}$ and $\vec{B}$ are such oriented that their vector product $\vec{E}\times\vec{B}$ points in the direction of propagation.

IMPEDANCE OF DIELECTRIC MEDIA

We can also write equation (1) in a more appropriate form as follows. Since $B_z = \mu H_z$ where μ is permeability of the medium,

$$\frac{E_y}{H_z} = \frac{\mu}{\sqrt{\mu\varepsilon}}$$

or $$\frac{E_y}{H_z} = \sqrt{\mu/\varepsilon} \qquad \text{...(1)}$$

However, for free space,

$$\frac{E_y}{H_z} = \sqrt{\frac{\mu_o}{\varepsilon_o}} \qquad \text{...(2)}$$

Dimensions of equation (2) expressed in MKS units are,

$$\frac{\text{volts / metre}}{\text{Amps. / metre}} = \frac{\text{volts / metre}}{\text{Amps. / metre}} = \text{ohms}$$

Thus $\sqrt{\mu/\varepsilon}$ has the dimensions of an impedance and the ratio E_y/H_z is called the impedance of the medium. For free space. $E_y/H_z = 376$ ohms.

EFFICIENCY OF A GENERATOR

Efficiency is defined as

$$\eta = \frac{\text{Electrical energy supplied by the generator}}{\text{Mechanical energy supplied to the generator}}$$

If V is terminal voltage developed and I is current in external circuit, then power developed = VI. If the total power loss is W watts, then power used

$$= (\text{VI} + \text{W}) \text{ watts}$$

and

$$\%\eta = \frac{\text{IV}}{\text{VI} + \text{W}} \times 100 \quad ...(1)$$

D.C. GENERATOR OR DYNAMO

IN d.c. dynamo the current and emf are unidirectional. The principle of working is same as that of a.c. generator except that an arrangement is provided which reverses the emf after half cycle in a.c. generator. This is achieved by reversing the connections to the ends of the coil of the generator, when the direction of emf changes every time. This is done by using a *split ring commutator* instead of the slip rings. The split ring consists of two segments S_1 and S_2 of a brass cylinder which are insulated from each other. Each end of the armature coil is connected to one segment of the split ring.

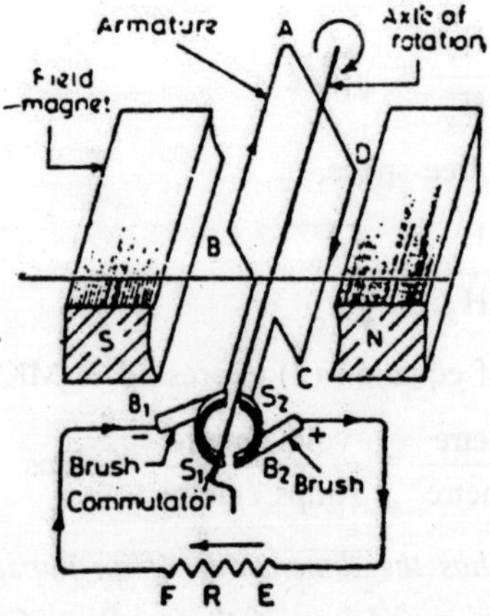

Fig. 3.2

The brushes and the split rings are so mounted that the contact of one brush with the first segment changes over to that with the second segment as soon as the direction of emf in the coil reverses. This is done by mounting the commutator on the same axis as the coil and rotating along with it. The emf reverses the sign when the plane of the coil is at right angles to the magnetic field $\vec{B}$ at this instant emf is zero. Such a split ring commutator is shown in Fig. 3.2.

The emf time graph is shown in Fig. 3.3.

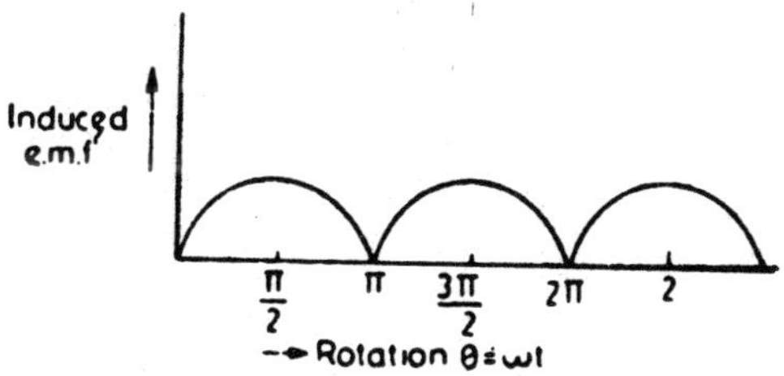

Fig. 3.3

ENERGY LOSSES IN GENERATOR AND THEIR REDUCTION

The energy losses in a dynamo are due to:

1. *Heating or copper losses.* Part of energy is wasted in overcoming the ohmic resistance of the armature coils, field coils and due to sparking at the brushes. This can be reduced by using thick copper wire of low resistance.
2. *Hysteresis Loss.* This is the energy used up in magnetising the core. This is reduced by choosing the material of the core having very thin B.H loop such as soft iron.
3. *Eddy Currents.* This is also known as iron loss and is due to eddy currents in the iron core. This is reduced by using laminated core.
4. *Flux Leakage.* Leakage of flux is reduced by making the pole-shoe large so as to enclose completely the armature and also by reducing the air gap.
5. *Mechanical Loss.* This is due to friction at the bearings, air resistance, etc. It is reduced by using roller or ball bearing and proper lubrication.

D.C. MOTORS

It is a device by which electrical energy is converted into mechanical energy. Hence, it is a reverse of the generator or dynamo.

Principle

The principle of a d.c. motor is the same as that of moving coil galvanometer which is the reverse of the d.c. generator. When a current carrying conductor capable of a free movement is placed in a magnetic field as shown in Fig. 3.4, it experiences a mechanical force and begins to move. When a coil carrying a current I is placed in a magnetic field $\vec{B}$, then it experiences a force $(\vec{F} = I\vec{l} \times \vec{B})$ which turns it about in a direction perpendicular to both the field and the current. Thus the coil rotates and the rotating torque

$$\tau = \vec{m} \times \vec{B}$$

where m is magnetic moment given by m = NIA, here A is area of the each coil and N are the number of the turns. When the plane of the coil is vertical, no torque acts on the armature (or coil). As the coil has sufficient inertia and frictional forces small it will cross the zero torque position.

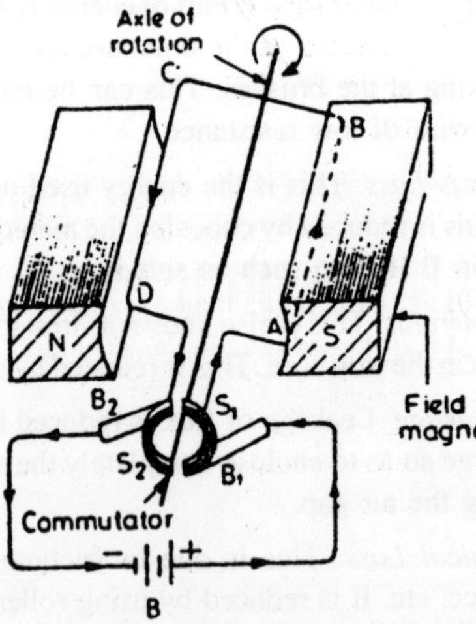

Fig. 3.4

In order to make the motion of the coil continuous, the direction of the current is reversed after the vertical position (*i.e.*, after each half revolution) with the help of split ring commutator so the torque acts in the same direction. Thus, we see that after each half cycle, there are positions of no torque and the principle used is reverse of d.c. generator. For smooth motion, the single coil is replaced by several coils set such as drum winding called armature.

The magnetic field is generally provided by electromagnets called field magnet. From the direct current sources, the armature is connected in series with electromagnet. Then motor is said to series-wound. If the armature is connected in parallel with the field magnet, the motor is then called parallel or shunt wound. Sometimes the windings over the armature and the field magnet are partly in series or partly in parallel, the motor is then called compound wound.

Back emf in a Motor

When the armature coil rotate an induced emf is set up in the coils of the winding and the direction of the induced emf opposes the current that rotates the armature. This emf is called back emf of the motor. Thus, every motor acts as a generator also.

The value of back emf goes on increasing steadily as the motor picks up speed. An equilibrium is established when this back emf plus the voltage drop due to the resistance of the armature coil is equal and opposite to potential applied to the coil.

Suppose applied potential difference $= V$

Back emf developed $= E$

Resistance of armature circuit $= R$

Current in armature $= I$

then,
$$V = E + IR \quad ...(1)$$

or
$$I = \frac{V - E}{R} \quad ...(2)$$

Thus, at the start when the speed is low, the back emf. is small so that armature current I is large. At full speed the back emf is large.

Starter

At start, sometimes the armature current is sufficiently large and it can damage the armature windings. To prevent this a variable resistance called started is connected in series with the armature when the speed increases, the starter resistance is reduced.

Efficiency of Motor

The input power used in the following ways,

1. Produces heat in the armature *i.e.*, heat loss I^2R
2. In magnetising the core of the armature is magnetic loss.
3. Mechanical power available.

If V is potential difference and I is current in armature, then,

Input power $= VI$

$$VI = I^2R + \text{magnetic loss} + \text{mechanical power}$$

If magnetic losses are negligible, we have,

$$VI = I^2R + \text{Mechanical power}$$

or mechanical power available,

$$= VI - I^2R$$

$$= I(V - IR), \qquad ...(3)$$

$$= IE,$$

$$E = V - IR = \text{back emf}$$

Efficiency is defined as the ratio of the output mechanical power to the electrical power

$$= \frac{\text{available mechanical power}}{\text{Input electrical power}}$$

$$= \frac{IE}{IV} = \frac{E}{V}$$

$$= \frac{\text{Back emf}}{\text{Applied emf}} \qquad ...(4)$$

This shows that efficiency increases as back emf E increases. Efficiency is maximum *i.e.*, 100 per cent,

When E = V, but from the relation E = V – IR, under this condition, I = 0, and motor will do no work because the current is zero, hence this cannot be realised in practice.

However, the efficiency is maximum when,

$$EI = \text{maximum}$$

or $$E\left(\frac{V-E}{R}\right) = \text{maximum, (from eq. 2)}$$

or $$\frac{EV}{R} - \frac{E^2}{R} = \text{maximum}$$

Differentiating with respect to E, we get,

$$\frac{V}{R} - \frac{2E}{R} = 0$$

or $$V = 2E$$

or $$E = \frac{V}{2} \qquad ...(5)$$

Hence for maximum efficiency, the back emf should be half the applied voltage. In actual practice this condition cannot be satisfied as the armature resistance being small, the armature current will become high and burn the coil of the armature. Therefore, the back emf in actual practice is allowed to go upto 90 to 95 per cent of the applied emf to keep armature current within the safe limit.

ENERGY DENSITY OF ELECTROMAGNETIC WAVE AND POYNTING THEOREM

One important characteristic of an electromagnetic wave is that it can transport energy from point to point. The direction of electromagnetic wave at a given point is the direction in which energy is being transmitted. We may describe a wave in terms of the new vector $\vec{S}$ which is defined as follows:

$$\vec{S} = \frac{1}{\mu_o}(\vec{E} \times \vec{B}) \text{ or } (\vec{E} \times \vec{H})$$

By this definition the direction of a wave is the direction of the vector $\vec{S}$ *called the "poynting vector"* after John Henry Poynting who first pointed out its properties. Let us now calculate the divergence of Poynting vector $\vec{S}$ for electromagnetic field in free space.

$$\nabla.\frac{1}{\mu_o}(\vec{E}\times\vec{B}) = \frac{1}{\mu_o}(\vec{B}.(\nabla \times \vec{E}) - \vec{E}.(\nabla \times \vec{B})$$

Substituting $\nabla \times \vec{E}$ and $\nabla \times \vec{B}$ from eq. (1.20), we get,

$$\nabla.\frac{1}{\mu_o}(\vec{E}\times\vec{B}) = \vec{E}\times\vec{B}\nabla.\frac{1}{\mu_o}\left(-\vec{B}.\frac{\delta\vec{B}}{\delta t} - \vec{E}_0.\mu_0\frac{\delta\vec{D}}{\delta t}\right)$$

$$= -\frac{1}{\mu_o}\left[\mu_o\varepsilon_o\vec{E}.\frac{\delta\vec{E}}{\delta t}+\vec{B}.\frac{\delta\vec{B}}{\delta t}\right] \text{ [since } \vec{D}=\varepsilon_o\vec{E}]$$

$$= -\frac{\delta}{\delta t}\left[\frac{1}{2}\varepsilon_o E^2+\frac{1}{2\mu_o}B^2\right]$$

or $$\nabla.(\vec{E}\times\vec{H}) = -\frac{\delta}{\delta t}\left[\frac{1}{2}\varepsilon_o E^2+\frac{1}{2}\mu_o H^2\right] \text{ (since B = moH)}$$

On integrating this over the volume v enclosed by the surface S, we get,

$$\int_v \nabla.(\vec{E}\times\vec{H})dv = -\frac{\delta}{\delta t}\int_v\left(\frac{1}{2}\varepsilon_o E^2+\frac{1}{2}\mu_o H^2\right)dv \quad ...(1)$$

The left hand side can be transferred to the surface integral by divergence theorem, we get,

$$\int_8 (\vec{E}\times\vec{H}).d\vec{A} = -\frac{\delta}{\delta t}\int_v\left(\frac{1}{2}\varepsilon_o E^2+\frac{1}{2}\mu_o H^2\right)dv \quad ...(2)$$

where $d\vec{A}$ is area.

For any other medium of permittivity e and permeability m, we have,

$$\int_8 (\vec{E}\times\vec{H}).d\vec{A} = -\frac{\delta}{\delta t}\int_v\left(\frac{1}{2}\varepsilon E^2+\frac{1}{2}\mu H^2\right)dv \quad ...(3)$$

As we know that electrostatic field energy per unit volume is $\frac{1}{2}\varepsilon E^2$ $\left(\text{or } \frac{1}{2}\varepsilon_0 E^2 \text{ for free space}\right)$ and the magnetostatic field energy per unit volume is $\frac{1}{2}\mu H^2$ $\left(\text{or } \frac{1}{2}\mu_0 H^2 \text{ for free space}\right)$. We shall assume that these expressions are valid even though the electric and magnetic fields vary rapidly. Thus the total electromagnetic energy in a given volume in free space is given by,

$$W = \frac{1}{2}\int_v(\varepsilon_0 E^2+\mu_0 H^2)dv \quad ...(4)$$

therefore from eq. (1.44a)

$$\int_8 (\vec{E}\times\vec{H})dA = -\frac{dW}{dt}$$

The term – dW/dt represents the rate at which energy is decreasing within the given volume. *The rate of decrease of energy must be equal to the net rate at which energy is flowing out of the volume. Therefore the left hand side represents the total outward flux of energy in watts over the surface enclosing the volume.* The $\vec{S}$ represents the amount of power per unit area. The integral of $\vec{S}$ over the closed surface clearly represents the rate at which energy leaves the enclosed volume.

Thus, for any arbitrary closed surface, the amount of power flowing out through the surface is

$$\oint \vec{S}.d\vec{A}$$

where $d\vec{A}$ is an element of area, therefore,

$$-\frac{dW}{dt} = \oint \vec{S}.d\vec{A} \qquad ...(5)$$

Equation (5) is known as "Poynting's theorem". It states an important property of $\vec{S}$ (the Poynting vector).

$$\vec{S} = \frac{1}{\mu_o}(\vec{E} \times \vec{B}) \text{ or } (\vec{E} \times \vec{H}) \text{ watts/metre}^2$$

represents the power flowing through a unit area normal to the direction of propagation is known as Poynting vector.

Be careful to note that i some situations $\vec{E} \times \vec{H}$ does not represent energy flow, as for example, in a static magnetic field superimposed on a static electric field. Hence *the interpretation that the Poynting vector is representing the flow of energy is not absolute or regorous rule but usually correct.*

GRASSOT'S FLUX METER

It is essentially a moving coil galvanometer in which mechanical damping and restoring torque are very much reduced but electromagnetic damping is relatively increased. Hence it is a modified moving coil ballistic galvanometer and is used to measure magnetic flux. The advantage of a Grassot's fluxmetre over a ballistic galvanometer is that while measuring the flux, the change in flux did not take place rapidly, since the fluxmetre coil is at rest before and after the change. Thus is Grassot's fluxmetre, the deflection depends only upon the change of flux and not upon the time taken for the flux to change, so the deflection is independent of the rate at with flux changes. The moving coil is suspended by a single silk fibre, the upper end of the fibre being attached to a flat spiral spring A. The coil is connected through two spirals B. B of thin silver foil and is suspended in the magnetic field of a permanent magnet NS. The search coil is connected to the terminals T_1 and T_2 of the moving coil of the fluxmetre as shown in Fig. 3.5.

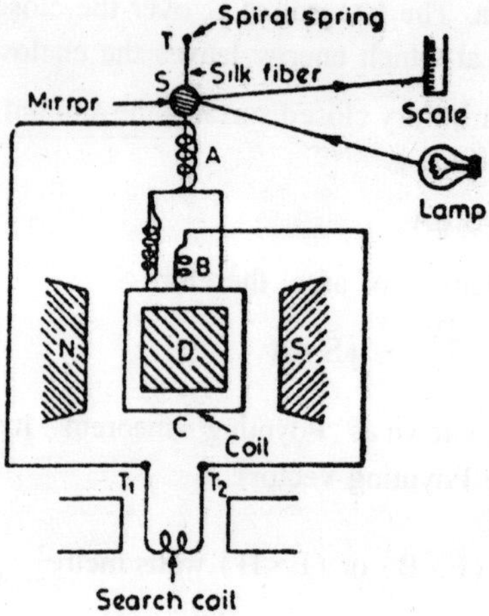

Fig. 3.5

The deflection of the coil is indicated by the usual lamp and scale arrangement. The search coil is inserted in the space where the magnetic flux is to be measured and then withdrawn, an emf is induced in the circuit which gives rise to a current. The corresponding deflection produced in the fluxmetre is noted.

The induced current = x/R

where x is induced emf and R is the resistance of the circuit

Suppose area of the fluxmetre coil = A

Number of turns = N

Magnetic field = B

The magnitude of the deflecting couple = iANB

(Since $\tau = \vec{m} \times \vec{B}$ and m = iA)

This couple acts on the fluxmetre coil. If i is the moment in inertia of the fluxmetre coil and $\frac{d\omega}{dt}$ is its angular acceleration then we have.

$$\text{NBiA} = I\ \frac{d\omega}{dt}$$

$$\text{ABN}\ \frac{\xi}{R} = I\ \frac{d\omega}{dt} \qquad ...(1)$$

The emf ξ is the resultant of two emfs one due to the rate of change of flux *i.e.*, $\frac{d\phi}{dt}$ in the search coil and the other due to the rotation of fluxmetre coil with angular velocity ω *i.e.*, NABω

$$\xi = \frac{d\phi}{dt} - \text{NAB}\ \omega$$

Substituting ξ in equation (3.15), we get,

$$\frac{\text{ABN}}{R}\left(\frac{d\phi}{dt} - \text{NAB}\ \omega\right) = I\ \frac{d\omega}{dt}$$

Integrating and putting $\omega = \frac{d\theta}{dt}$

$$\frac{\text{NAB}}{R}\left(\int_0^t \frac{d\phi}{dt}\,dt - \text{NAB}\int_0^t \frac{d\theta}{dt}\,dt\right) = I\int_0^t \frac{d\omega}{dt}\,dt$$

The integral of the right-hand side *i.e.*,

$$\int_0^t d\omega = |\omega|_{t=0}^{t=t} = 0$$

because the coil is at rest before and after the change of flux *i.e.*, angular velocity is zero when t = 0 and t = t.

$$\frac{\text{NAB}}{R}\left[\int_0^t d\phi - \text{NAB}\int_0^t d\theta\right] = 0$$

or $$\int_0^t d\phi - \text{NAB}\int_0^t d\theta = 0$$

or $$\phi - \text{NAB}\phi = 0$$

or $$\phi = \text{NAB}\ \theta$$

or $$\phi = K\ \theta \qquad ...(2)$$

where $K = \text{NAB}$.

Thus, *the total change in flux in the search coil is independent of time in which the change of flux takes place.* The deflection θ in the fluxmetre is proportional to the flux linkage in the search coil. Therefore, the displacement of the fluxmetre coil, when the search coil is inserted in or withdrawn from the unknown field, measures the flux linked with the search coil.

A.C GENERATOR OR DYNAMO

Principle

It is based on the principle of electromagnetic induction. When a coil rotates in a uniform magnetic field, the associated flux changes and an induced emf is generated according to Faraday's law of electromagnetic induction. *Thus the rotation of coil (Mechanical Processes) produces induced emf (electrical effects) if the process is made continuous, then we can convert continuously mechanical energy into electrical energy.*

Suppose a coil of wire is rotated about an axis OO' in a uniform magnetic field $\vec{B}$. The flux passing through the coil.

$$\phi = \int_S \vec{B}.\hat{n}dS$$

where $\hat{n}$ is a vector along the outward normal to the plane of the coil and dS is an element of the area enclosed by the coil. Let at any instant θ be the angle between normal to the plane of the coil and $\vec{B}$. Further, let N be the total number of turns in the coil and A be the area of each turn, total flux

$$\phi = \int_S B \cos\theta \, dS$$

$$= B\cos\theta \int_S dS$$

$$= NAB\cos\theta, \text{ since } \int_S dS = NA$$

The induced emf in the coil at any instant t is given by,

$$E = -\frac{d\phi}{dt} = -\frac{d}{dt}(BAN\cos\theta)$$

$$= BAN\sin\theta\frac{d\theta}{dt}$$

If ω is the angular velocity, then ωt = θ, therefore,

$$E = BAN\,\omega\sin\omega t \qquad ...(1)$$

The value of induced emf is maximum when $\omega t = \frac{\pi}{2}$, $\sin \omega t = 1$, *i.e.*, when the plane of the coil is parallel to the magnetic field, this maximum value of E is denoted by E_o.

$$E_o = BAN\omega$$

Equation (1) becomes,

$$E = E_o \sin \omega t \qquad ...(2)$$

This relation (2) shows the variation of emf with time.

1. When $\omega t = 0$, *i.e.*, the plane of the coil is perpendicular to the field (coil is vertical), $E = 0$.
2. When $\omega t = \frac{\pi}{2}$, *i.e.*, the plane of the coil is parallel to B (coil is horizontal) we have $E = E_o$.
3. When $\omega t = \pi$, *i.e.*, the plane of the coil is again perpendicular to the field, $E = 0$.
4. When $\omega t = \frac{3\pi}{2}$, *i.e.*, the plane of the coil is again parallel to the field $\sin\left(\pi + \frac{\pi}{2}\right) = \sin \frac{\pi}{2} = -1$, so that $E = -E_o$.

   It shows that the direction of the induced emf is again maximum but in the reverse direction.
5. When $\omega t = 2\pi$, the plane of the coil is again perpendicular to the field and $E = 0$.

Thus, the emf induced in the coil goes through a cycle. Such emf is called sinusoidal just like sine wave. It is also referred to as alternating emf. The corresponding current in the external circuit,

$$I = \frac{E}{R} = \frac{E_o}{R} \sin \omega t$$

$$= I_o \sin \omega t \qquad ...(3)$$

where I_o is maximum current.

Construction

A simple electric generator consists of a coil which can rotate in magnetic field. The coil of the wire is called the armature. It is held between the pole-pieces of strong magnet called the field magnet. When the armature rotates, the magnetic flux linked with it changes and the current is induced in the coil. This current is taken out by means of an arrangement called slip rings and brushes. The two rings are insulated from each other the brush lightly touches each ring and connects the armature with the external circuit.

Source of Energy

The emf is generated when the armature rotates, hence, there must by some device for rotating the armature. Such as in a cycle dynamo the armature is rotated by the rotation of the wheel. For other purposes, steam engines, gas turbines etc. are used to rotate the armature. Thus the mechanical energy is converted into electrical energy. *In every ;generator, some form of mechanical energy is put into the generator, and transformed into electrical energy. Therefore the generator basically does not produce energy but merely it transforms mechanical to electrical forms.*

ENERGY PER UNIT VOLUME

In Fig 3.6 an area A is normal to the x-axis if a plane wave is travelling in the positive x-direction with speed v, then all the energy contained in the volume A v dt will pass through the area A in time dt.

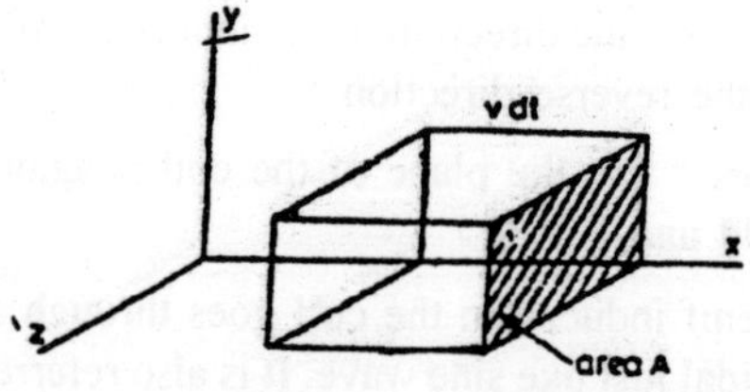

Fig. 3.6

Thus, if U is energy per unit volume, the total energy contained in the volume A v dt is U A v dt. Therefore the energy passing through unit area per unit time is,

$$S = \frac{UA\,v\,dt}{A\,dt} = Uv$$

Thus, $\quad U = S/v \qquad ...(1)$

MEASUREMENT OF A STRONG MAGNETIC FIELD USING A SEARCH COIL

A search coil is a solenoid of small diameter having few (about 50) turns of insulated copper wire and is suitable for measuring strong magnetic field.

Suppose we have to find the magnetic field between the pole-pieces of an electromagnet. The search coil is placed in the gap between the pole pieces and is connected in series with a ballistic galvanometer through a high resistance box R to control the throw in B.G. and the secondary coil SS of standard solenoid as shown in Fig. 3.7 We can change the magnetic flux in two ways.

(1) Put on the magnetic field of the electromagnet and the search coil is either suddenly inserted in or withdrawn from the field, *i.e.*, the region between pole pieces NS, with its plane at right angles to the magnetic field.

(2) The search coil is held between the pole pieces and the magnetic field is suddenly put on.

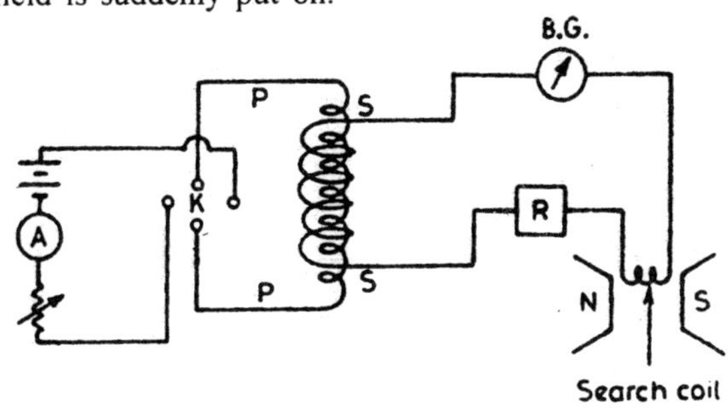

Fig. 3.7

In both cases the flux linked with the search coil changes and produces an induced emf in the circuit which gives throw in ballistic galvanometer.

Suppose the throw	$= \theta$
Number of turns of the search coil	$= N$
The area of each turn of the coil	$= A$
The magnetic field (or flux density)	$= B$
Then flux linked with the search coil	$= NAB$

But this represents change in flux in both cases mentioned above, since in the first case initially the flux was zero whether the coil was outside or inside, as the case may be. In the second case mentioned above the flux becomes zero when the field is put off. Thus in both cases,

The change in flux $d\phi = NAB$

Induced emf in time $dt = \dfrac{NAB}{dt}$

$$\text{Current} \quad I = \frac{NAB}{Rdt}$$

The charge flowing in time dt

$$q = Idt = \frac{NAB}{R}$$

This charge flows through B.G. them,

$$\frac{NAB}{R} = K\theta\left(1 + \frac{\lambda}{2}\right) \qquad ...(1)$$

In order to determine k, *i.e.*, for calibrating the galvanometer a series circuit containing a battery, an ammeter and a rheostat is connected to the primary of the standard solenoid. A known current in the primary circuit is passed. It produces the field inside the solenoid $\mu_o n_p I$. This causes a magnetic flux to pass through the secondary. The flux linked with secondary coil = $\mu_o n_p N_s A_s I$.

where n_p are the number of turns per unit length in the primary coil,

N_s the total number of turns in the secondary coil

A_s the area of each secondary coil and I is the current through the primary.

Now, the direction of current in the primary is suddenly reversed with reversing key, the total change in flux,

$$= \mu_o n_p N_s A_s I - (\square\, \mu_o n_p N_s A_s I)$$

$$= 2\,\mu_o n_p N_s A_s I$$

As above, the charge due to this change in flux

$$= \frac{2\,\mu_o\ n_p\ N_s\ A_s\ I}{R}$$

Let θ_1 be the throw observed in ballistic galvanometer when the charge passes through it then,

$$= \frac{2\,\mu_o\, n_p\, N_s\, A_s\, I}{R} = k\,\theta_1 \left(1+\frac{\lambda}{2}\right) \qquad ...(2)$$

Dividing eqn. (1) by (2), we get,

$$= \frac{NAB}{2\,\mu_o\, n_p\, N_s\, A_s\, I} = \frac{\theta}{\theta_1}$$

or $$B = \frac{2\,\mu_o\, n_p\, N_s\, A_s\, I}{NA} = \frac{\theta}{\theta_1} \text{ weber / metre}^2 \qquad ...(3)$$

Hence, B can be determined.

While calibrating the galvanometer, if instead of reversing the direction of current in the primary, the current is suddenly started or stopped, the change in flux linked with secondary will be $\mu_o n_p N_s A_s I$ and the throw θ_2 in B.G. will now be half of the θ_1, thus,

$$B = \frac{2\,\mu_o\, n_p\, N_s\, A_s\, I}{NA} = \frac{\theta}{\theta_2} \qquad ...(4)$$

Hence, B can be determined.

BASIC EQUATIONS

The basic equations of electricity and magnetism which we have studied are summarised in differential form by the following four equations.

(1) Gauss's law for electric field of charge, *i.e.*,

$$\text{div } \vec{D} = \nabla . \vec{D} = r \qquad ...(1)$$

(2) Gauss's law for magnetic field *i.e.*,

$$\text{div } \vec{B} = \nabla . \vec{B} = 0, \qquad ...(2)$$

(3) Faraday's law for induced emf produced by the rate of change of magnetic flux, *i.e.*,

$$\text{Curl } \vec{E} = \nabla \times \vec{E} = -\frac{\delta \vec{B}}{\delta t}, \qquad ...(3)$$

(4) Ampere's law for magnetic field due to Transformers and A.C. Bridges *i.e.*,

$$\text{Curl } \vec{B} = \nabla \times \vec{B} = \mu_o \vec{J} \qquad ...(4)$$

The first three equations are valid for static as well as dynamic field and they apply in general. However, the fourth law *i.e., Ampere's law was derived for Transformers and A.C. Bridges and does not hold for time varying field.*

TYPES OF CURRENTS

We shall the discuss currents in material in a manner similar to that used to treat charges in material medium. In general, currents may be classified in two categories:

(1) *True currents* due to the motion of true charges.

(2) *Other currents* which are due to medium itself.

This separation is analogous to the separation that was discussed in the electrostatic theory between the potential (or fields) of true charges and the potentials of polarisation charges. Similarly here, there are two types of magnetic fields, one derived from the true currents and the other derived from the combined effects of all the currents whatever may be their origin. It is this latter field, namely, the field of magnetic induction $\vec{B}$, that can be considered to be space-time average of the interatomic fields.

All currents are classified as follows:

(1) *True currents* indentical with the physical transport of true charges.

(2) *Polarisation currents*. Currents those arise from the change of polarisation with time.

(3) *Magnetisation currents*. These are stationary currents that flow within regions that are inaccessible to observation but might give rise to net boundary or volume currents, due to imperfect orbit cancellation on an atomic scale.

(4) *Convective currents*. If a material medium is in motion and contains charges of various types. Additional currents are obtained those arise from the motion of both true and polarisation charges contained in the medium. Such currents are called convective currents.

Therefore in a stationary medium the total current is given by the sum of the first three types of currents given above. Thus,

$$\vec{J}_{total} = \vec{J}_{true} + \frac{\delta \vec{P}}{\delta t} + \nabla \times \vec{M} \qquad ...(1)$$

VACUUM DISPLACEMENT CURRENT

Taking the divergence of equation (1), we get,

$$\text{div curl } \vec{B} = \mu_0 \text{ div } \vec{J}$$

The left hand side will become zero. *Since the divergence of curl of a vector is always zero.* Hence, the right hand side is also zero,

$$\text{div } \vec{J} = \nabla.\vec{J} = 0 \qquad ...(1)$$

Here $\vec{J}$ is total current as given in equation (1).

This means that the total flux of current out of nay closed surface is zero. Physically, this means that current is always closed and there are no sources or sinks. Thus we arrive at a contradiction. Since for time varying case *i.e.,* for changing electric fields such as the discharge of a charged condenser through a conducting wire joining its plates. The current starts at positively charged plate, whose charge gradually decreases as the current flows through the conducting wire to the negatively charged plate and neutralises the negative charge on it. Thus the condenser plates are regarded as sources or sinks. *Hence the Ampere's law is valid only for steady state phenomena and not for changing fields.*

Equation (2) is also in conflict with the equation of continuity which gives the complete relation

$$\text{div } \vec{J} = \frac{\delta\rho}{\delta t} \qquad ...(2)$$

Maxwell concluded that equation (1) is incomplete. The contradiction of eq. (1) with the equation of continuity *i.e.,* (2) could be removed by converting the equation of continuity into a form whose divergence is zero by adding something to $\vec{j}$ in equation (2). Hence to assume the conservation of charge, it is necessary that total current should obey the equation of continuity

Gauss's law is

$$\nabla.\vec{D} = \rho$$

Differentiating with respect to time

$$\frac{\delta}{\delta t}\nabla.\vec{D} = \frac{\delta\rho}{\delta t}$$

$$\nabla.\frac{\delta\vec{D}}{\delta t} = \frac{\delta\rho}{\delta t}$$

Adding $\nabla.\vec{J}$ on both sides,

$$\nabla.\vec{J} + \frac{\delta\rho}{\delta t} = \nabla.\vec{J} + \nabla.\frac{\delta\vec{D}}{\delta t}$$

$$= \nabla.\left(\vec{J} + \frac{\delta\vec{D}}{\delta t}\right)$$

where $\vec{D} = \varepsilon_o\vec{E}$ is displacement vector in free space. According to equation of continuity,

$$\nabla.\vec{J} + \frac{\delta\rho}{\delta t} = \nabla.\left(\vec{J} + \frac{\delta\vec{D}}{\delta t}\right) = 0$$

Thus $\nabla.\vec{J} = 0$, eq. (1.6) for Transformers and A.C. Bridges

and $\nabla.\left(\vec{J} + \frac{\delta\vec{D}}{\delta t}\right) = 0$, every where ...(3)

i.e., for steady currents as well as currents corresponding to time varying fields.

So that Maxwell replaced $\vec{J}$ in Ampere's law by,

$\vec{J} + \frac{\delta\vec{D}}{\delta t}$ and Ampere's law becomes

$$\text{curl } \vec{B} = \mu_o\left(\vec{J} + \frac{\delta\vec{D}}{\delta t}\right) \quad ...(4)$$

The term $\frac{\delta\vec{D}}{\delta t}$ is called is the *displacement current density* since it arises when the electric displacement $\vec{D}$ (*i.e.,* in effect $\vec{E}$) changes with time. Maxwell assumed that it is just as effective as $\vec{J}$ in producing magnetic field. The introduction of displacement current is one of Maxwell major contribution to the electromagnetic theory without which the, electromagnetic theory would have been impossible.

In fact, relation (4) is called Maxwell's equation as derived from Ampere's law.

Significance of Displacement Current

The vacuum displacement current $\varepsilon_0 \dfrac{\delta E}{\delta t}$ does not have the significance of a current in the sense of being the motion of charges, with the introduction of displacement of current the total current given in equation (5) is,

$$\vec{C} = \vec{J}_{true} + \frac{\delta \vec{P}}{\delta t} + \nabla \times \vec{M} + \varepsilon_0 \frac{\delta \vec{E}}{\delta t}$$

$$\vec{C} = \vec{J}_{total} + \varepsilon_0 \frac{\delta \vec{E}}{\delta t} \qquad ...(5)$$

The current C obtained by adding the term $\varepsilon_0 \dfrac{\delta \vec{E}}{\delta t}$ to the total current, is a solenoidal current, of which divergence is zero. The introduction of displacement current is necessary in order to apply the Ampere's law for magnetic field due to non-stationary currents. The geometrical significance of the solenoidal current C is that at points where there is an accumulation of charge, the current is assumed to be continuous across the discontinuity in the form of the rate of change of the field resulting from the accumulation of charge on the boundaries of the discontinuity. As an example charging a capacitor produces a closed current loop in terms of C.

Examples of Displacement

The idea of displacement current originated from the study of discharge of condenser. Before Maxwell the discharge current was supposed to flow in the open circuit formed by the condenser plates and the wires. Maxwell interpreted that all electric currents are divergenceless *i.e.,* electric currents must be closed and in the discharge of condenser the conduction current is closed by displacement current (arising due to change of electric displacement in the dielectric) flowing in the dielectric between the plates. Consider a voltage applied to a resistor and a capacitor in parallel as shown in Fig. 3.8.

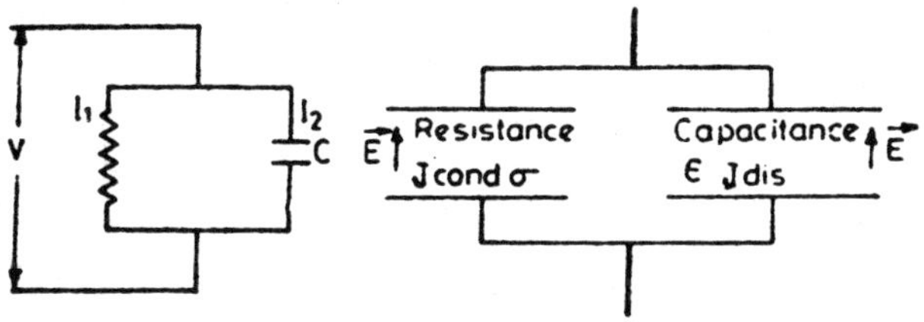

Fig. 3.8

The nature of current flowing through the resistor is different from that through the capacitor. A constant voltage across a resistor produces a continuous flow of current of constant value, on the other hand, the current through a capacitor will be constant only when the voltage is changing.

The current I_1 through the resistor is given by,

$$I_1 = \frac{V}{R}$$

and the current through the capacitor is given by,

$$I_2 = \frac{dq}{dt} = C\frac{dV}{dt}$$

Since the instantaneous charge in the capacitor is given by q = CV.

The current through the resistor is a conduction current, while the current through the capacitor is displacement current. Although, the current does not flow through the capacitor, the external effect is as though it did. Since as much current flows out of one plate as flows into the opposite one.

The electric field $\vec{E}$ equals the voltage V across the element divided by its length d, that is

$$E = V/d$$

The current density J_1 inside the resistor equals the product of the electric field E and the conductivity of the medium inside the resistor element and also I_1 divided by the cross-sectional area A or

$$J_1 = \sigma E = \frac{I_1}{A}$$

The capacitance of parallel plate capacitor

$C = \frac{\varepsilon A}{d}$, where ε is permittivity of the medium, then

$$I_2 = \frac{\varepsilon A}{d}.\frac{d}{dt}(Ed) = \varepsilon A\frac{dE}{dt}$$

and current density

$$J_2 = \frac{I_2}{A} = \frac{\in dE}{dt}$$

Recalling that D = εE

$$J_2 = \frac{dD}{dt}$$

In this example J_1 is a conduction current density J_{cond}, while J_2 is a displacement current density J_{disp}. Since the current density $\vec{J}$, the electric displacement $\vec{D}$ and the electric field intensity $\vec{E}$ are actually space vectors which all have the same direction in isotropic media, hence we can write,

$$\vec{J}_{cond} = \sigma\vec{E}$$

and $$\vec{J}_{disp} = \frac{\varepsilon d\vec{E}}{dt} = \frac{d\vec{D}}{dt}$$

The total current density $\vec{J}_{total}$

$$\vec{J}_{total} = \vec{J}_{cond} + \vec{J}_{disp} \qquad ...(6)$$

Thus there is a displacement current density in the region of condenser plates, *the concept of displacement current has very important consequences for dielectrics and for free space.* The displacement current is responsible for the production of magnetic fields in empty space since the conduction current is zero and the magnetic fields are entirely due to displacement currents. The displacement current cannot be detected directly but the confirmation of its existence is provided indirectly by the electromagnetic theory.

CONDITION WHEN A MOVING COIL GALVANOMETER IS DEAD BEAT OR BALLISTIC

In deriving equation (1) we have assumed that whole of the kinetic energy of the coil is used up in twisting the fibre and neglected the retarding forces present namely, (1) due to friction of air, (2) viscosity of suspension fibre and (3) induced current in any neighbouring mass or metal. The first two of these factors constitute mechanical damping, while the third factor accounts for electromagnetic damping. The retarding couple due to damping varies as the angular velocity = dθ/dt, where p is coefficient of mechanical damping.

When the circuit is closed, we have in addition the induced current produced in the coil which varies as the rate of change of displacement and inversely as the total resistance of the circuit, hence the retarding couple is $\frac{m}{R}\frac{d\theta}{dt}$, where the constant m involves the magnetic flux due to the magnet and the area of coil. Thus, the total retarding couple due to damping

$$= \left(p + \frac{m}{R}\right)\frac{d\theta}{dt}.$$

Now the angular acceleration is $\frac{d^2\theta}{dt^2}$.

Hence, according to Newton's second law, equation of motion is,

$$\frac{Id^2\theta}{dt^2} = -c\theta - \left(p + \frac{m}{R}\right)\frac{d\theta}{dt}$$

or
$$\frac{Id^2\theta}{dt^2} + \left(\frac{m}{R} + p\right)\frac{d\theta}{dt} + c\theta = 0$$

or
$$\frac{d^2\theta}{dt^2} + \frac{1}{I}\left(\frac{m}{R} + p\right)\frac{d\theta}{dt} + \frac{c\theta}{I} = 0$$

or
$$\frac{d^2\theta}{dt^2} + 2b\frac{d\theta}{dt} + K^2\theta = 0 \quad \text{...(1)}$$

where,
$$2b = \frac{1}{I}\left(\frac{m}{R} + p\right) \quad \text{...(2)}$$

and $K^2 = \frac{c}{I}$ ...(3)

Let the solution of this equation be $\theta = Ae^{\alpha t}$

Then,
$$\frac{d\theta}{dt} = A\alpha e^{\alpha t} \text{ and } \frac{d^2\theta}{dt^2} = A\alpha^2 e^{\alpha t}.$$

Putting these values in equation (1),

$$\alpha^2 + 2b\alpha + K^2 = 0$$

or
$$\alpha = -b \pm \sqrt{b^2 - K^2}$$

This gives,

$$\theta = A_1 e^{(-b + \sqrt{b^2 - K^2})t} + A_2 e^{(-b - \sqrt{b^2 - K^2})t}$$

where, A_2 and A_1 are constants.

The quantity $b^2 - K^2$ may be positive, zero or negative depending upon whether $b > K$, $b = K$ or $b < K$.

Case 1: When $b > K$, we have two real values of a, then solution of equation (1) is,

$$\theta = A_1 e^{\alpha 1t} + A_2 e^{\alpha 2t}$$

The α_1 and α_2 are negative. The deflection θ will go no decreasing with increasing time. *The motion is non-oscillating or dead beat and the oscillation is said to be over-damped.*

Case 2 : Critical damping. When $b = K$, the two values of a are equal to $- b$ and solution of (1) becomes

$$\theta = (A_1 + A_2)\, e^{-bt}$$

$$= A_3\, e^{-bt}$$

That the coil after deflection comes to rest in minimum possible time and the galvanometer is said to be *critically damped. However, the motion of the coil is just non-oscillatory*

Case 3 : When $b > K$, the quantity $\sqrt{b^2 - K^2}$ is imaginary so the values of a are complex *i.e.*,

$$\alpha = - b \pm jg. \text{ where } g^2 = K^2 - b^2$$

Hence, solution of equation (1) becomes,

$$\theta = A_1\, e^{(-b + jg)t} + A_2 e^{(-b - jg)t}$$

$$= e^{-bt}\, [A_1 e^{jgt} + A_2 e^{-jgt}]$$

$$= e^{-bt}\, [A_1 \{\cos gt + j \sin gt\}$$

$$+ A_2 \{\cos gt - j \sin gt\}]$$

$$= e^{-bt}\, [(A_1 + A_2) \cos gt + j\, (A_1 - A_2) \sin gt]$$

Putting $A_1 + A_2 = B \sin \beta$

and $j\,(A_1 - A_2) = B \cos \beta$

$$\theta = Be^{-bt} \sin (gt + \beta) = Be^{-bt} \sin \{(\sqrt{K^2 - b^2})\, t + \beta\}.$$

The motion is oscillatory or ballistic and period of oscillation,

$$T = \frac{2\pi}{g} = \frac{2\pi}{\sqrt{K^2 - b^2}} \qquad ...(4)$$

In order that the whole charge may pass through the ballistic galvanometer before it has moved from its zero position, it is essential that its periodic time be large. Since we have,

$$T = 2\pi \sqrt{\frac{I}{c}}.$$

therefore, I should be large and c should be small. As we have

$$b < K$$

$$\frac{1}{2I}\left(\frac{m}{R}+p\right) < \frac{c}{I} < \frac{2\pi}{T}$$

Hence to make b small,

I should be large,

R should be large,

m should be small and so the coil should be wound round a nonconducting frame, *e.g.*, paper, bamboo or plastic. p should be small, and so the air resistance should be small and the fibre should be very fine.

We just see that the requirements of dead-beat galvanometer are just opposite to those for the ballistic b > K. Hence the opposite conditions should prevail. The coil is made of thick copper wire and has smaller number of turns so to have smaller resistance and is wound round a metallic or conducting frame.

THE RETARDED POTENTIALS

We have shown that Maxwell's equations have wave type solutions and that the velocity of propagation is C in free space. The plane waves have been treated without inquiring how these waves were produced. We have not shown, however when these solutions are physically relevant or meaningful, nor how they are related to the charges and the currents that must be their sources.

Hence our next step is to consider prescribed charge and current distributions, ρ (r, t) and $\vec{J}$ (r, t) and find the fields produced by them that answer above questions. There are several ways of approaching the problem, of which the most fruitful is the potential approach, which is developed analogously to the procedures used in electrostatics and magnetostatics.

Therefore if we which to relate the fields of the changing charge and current distribution which produce them it is far easier to work with the potentials than with the fields directly. To be more specific, we shall discuss the generation of electromagnetic waves. We will determine the potentials in terms of given charge and current distribution. Once the potentials have been determined the fields can be obtained from the relation given in Eqs. (1) and (2).

Let the point P on Fig. 3.9, have a position vector r and the point Q have position vector r'. If the charge distribution near Q changes with time, the information that it has changed only be appreciated at the point P after the time r – r'/C. This is the time taken by electromagnetic radiation to travel the distance r – r'.

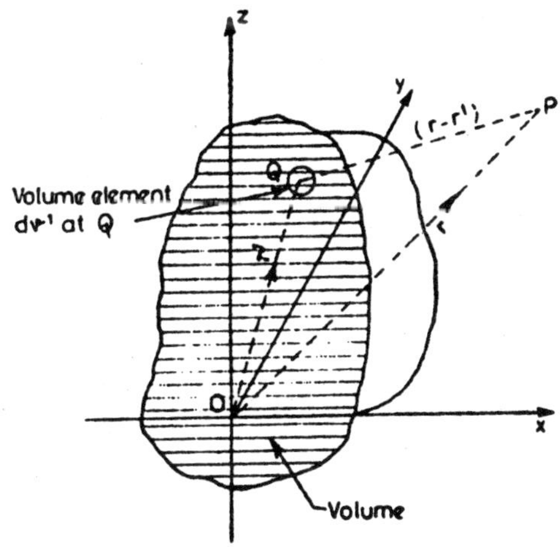

Fig. 3.9

It means that at a time t the contribution to the potential V(r, t) at P from the charge within the volume element dv' at Q depends not on what the charge is at time t but on what it was at the time $t - \frac{|r - r'|}{C}$. This is true for all he volume elements within the considered volume. Note that the time lapse $\frac{r - r'}{C}$ depends on the position vector r' and so varies over the total volume. The potential is given by

$$V(r, t) = \frac{1}{4\pi\varepsilon_0} \int_{\text{all}} \frac{\rho\left(r', t - \frac{|r - r'|}{C}\right)}{|r - r'|} dv' \qquad ...(1)$$

which is known as the "retarded scalar potential".

EARTH INDUCTOR

It is an important device based on electromagnetic induction. It consists of a circular coil C (Fig. 3.10) of 1000 to 5000 turns of insulated copper wire wound found a wooden frame of about 25 cm diameter. The circular frame is capable of rotation about an axis passing through two diametrically opposite points. There is a spring key which allows the coil to rotate through 180° only. The two ends of coil are connected to two terminals T_1 and T_2. They are sometimes also joined to a ballistic galvanometer through a commutator. The handle is used to bring the coil to its position so that on pressing the release key the coil is set in rotation.

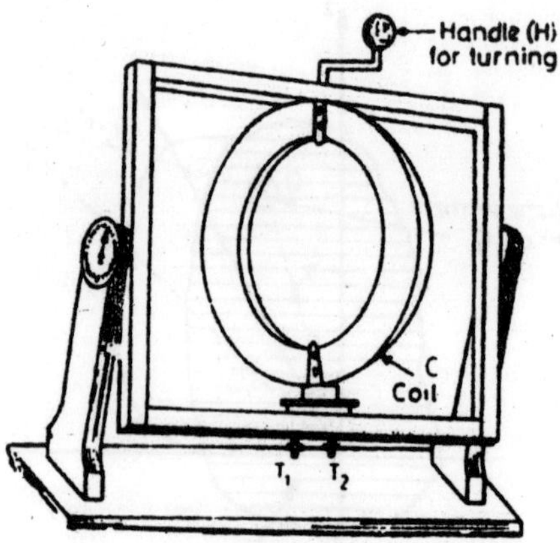

Fig. 3.10

Theory

Consider, a coil of n turns and area A is placed in a magnetic field B. When this coil is rotated in this uniform magnetic field, the magnetic flux associated with it changes and according to Faraday's law an induced emf is set up in it. Suppose at any instant the plane of coil is at right angles to the field, then flux linked with coil = nBA.

Suppose at any instant it makes an angle θ with a direction at right angles to that of the coil (Fig. 4.10) then the flux linked with it nBA cos θ.

Since flux = normal component × area

The induced emf,

$$\xi = -\frac{d\phi}{dt} = -\frac{d}{dt}(nBA\cos\theta)$$

$$= nBA \sin\theta \frac{d\theta}{dt}$$

If R is the resistance of the circuit, induced current is given by,

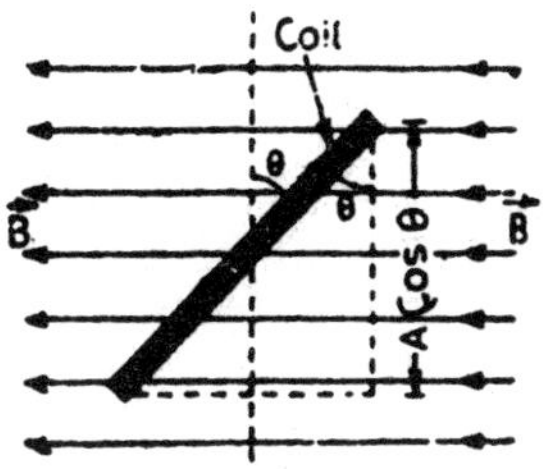

Fig. 3.11

$$I = \frac{\xi}{R} = \frac{nBA}{R} \sin\theta \frac{d\theta}{dt}$$

Therefore, the charge that flows through B.G. during a small interval of time dt is,

$$dq = Idt$$

Substituting I,

$$dq = \frac{nBA}{R} \sin\theta \, d\theta$$

when the coil is rotated through half the complete rotation *i.e.*, 0 to 180°, then the total charge passing through the ballistic galvanometer is given by,

$$\int dq = \int_0^{\pi} \frac{nBA}{R} \sin\theta \, d\theta$$

$$q = \frac{nBA}{R} [-\cos\theta]_0^{\pi}$$

$$= -\frac{nBA}{R} [\cos\pi - \cos 0]$$

$$= \frac{2nBA}{R} \quad \text{...(1)}$$

If θ_1 is the corresponding first throw in the ballistic galvanometer,

$$\frac{2nBA}{R} = k\theta_1 \left(1 + \frac{\lambda}{2}\right) \quad \text{...(2)}$$

where K is the ballistic constant and λ is the logarithmic constant.

STANDARD MAGNETIC FLUX

Standard magnetic fluxes are used for calibrating ballistic galvanometer. The earth inductor also works as a source standard flux is discussed below.

Standard Solenoid or Flux Meter

It consists of along solenoid of large number of turns as primary coil PP. The secondary coil SS of few turns wound over primary coil as shown in Fig. 3.12. Suppose the number of turns per unit length of the primary be n_p. When a steady current I is sent through the primary. The magnetic field the solenoid is given by,

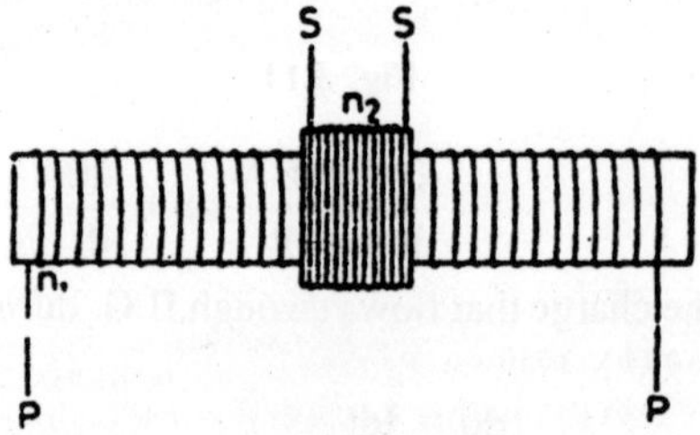

Fig. 3.12

$$B = \mu_o n_p I$$

Now, if A_s be the mean area of the secondary coil and N_s the total number of turns in it. As the lines of magnetic field $\vec{B}$ inside the solenoid also pass through the secondary, producing the magnetic flux through the secondary. The total flux through the secondary is,

$$\phi = BA_sN_s$$

The magnitude of the induced emf

$$\xi = \frac{d\phi}{dt} = \mu_o npN_sA_s \frac{dI}{dt}$$

Induced current $= \frac{\xi}{R} = \frac{\mu_o npN_sA_s}{R} \frac{dI}{dt}$

where R is the total resistance of the secondary,

And the charge q is given by

$$q = \int_0^I \frac{\mu_o npN_sA_s}{R} \frac{dI}{dt} dt$$

$$= \frac{\mu_0 npN_sA_sI}{R}$$

If θ is the observed throw in B.G., when this charge passes through it, then

$$\frac{\mu_0 npN_sA_sI}{R} = k\theta_H\left(1+\frac{\lambda}{2}\right) \qquad (1)$$

This equation is used for determining the ballistic constant K. Thus, the galvanometer is calibrated.

USES OF EARTH INDUCTOR

Measurement of Horizontal Component H of the Earth Magnetic Field

To measure the horizontal component of the earth's magnetic field at any place, the coil is set with its axis of rotation vertical and the plane of the coil east-west, *i.e.*, perpendicular to the magnetic meridian. The coil is now rapidly rotated through 180° by spring key, consequently induced charge due to change in flux flows through the B.G. which gives a deflection θ_H. From equation (1), we get

$$\frac{2NAH}{R} = k\,\theta_H\left(1+\frac{\lambda}{2}\right) \qquad (1)$$

where R is the total resistance and the coil rotates in field H, *i.e.*, the horizontal component of earth's magnetic field. If the value of K is known, we can determine H. K is calculated by calibrating B.G.

K can be calculated by sending a known current through the primary of standard solenoid and nothing the observed ballistic throw in the galvanometer. The circuit used for this purpose is shown in Fig. 4.12. Let the current in the primary be adjusted to I and the direction of the current is suddenly reversed.

The flux linked with the secondary will change by an amount, given by,

$$\mu_o N_p N_s A_s I - (\mu_o N_p N_s A_s I) = 2\mu_o N_p N_s A_s I$$

charge due to this change in flux,

$$q = \frac{2\mu_0 N_p N_s A_s I}{R}$$

If θ is the ballistic throw when this charge passes through ballistic galvanometer,

$$\frac{2\mu_o N_p N_s A_s I}{R} = k\theta\left(1+\frac{\lambda}{2}\right) \quad ...(2)$$

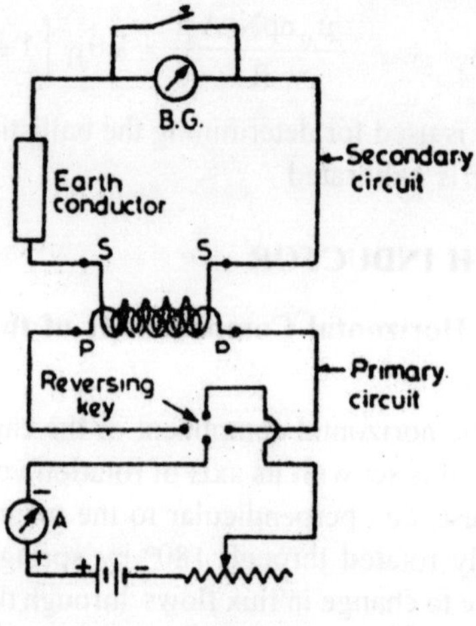

Fig. 3.13

Dividing equation (3.4) by (3.5),

$$\frac{NAH}{\mu_o N_p N_s A_s I} = \frac{\theta_H}{\theta}$$

or $$H = \frac{\mu_o N_p N_s A_s I}{NA}\frac{\theta_H}{\theta} \text{ weber / metre}^2 \quad ...(3)$$

Thus, H the horizontal component of the earth magnetic field is known in terms of know quantities.

Measurement of Vertical Component V of the Earth Magnetic Field

Here, the coil is adjusted so that is plane is horizontal and it can rotate about a horizontal axis in the magnetic meridian. The coil now rotates in the vertical component V of the earth's magnetic field. The coil is now rapidly rotated through 180° and the deflection θ in the ballistic galvanometer is observed. Proceeding exactly as in previous case, a relation for V is obtained *i.e.*,

$$V = \frac{\mu_o N_p N_s A_s I}{NA}\frac{\theta_v}{\theta} \text{ weber / metre}^2 \quad ...(4)$$

The intensity of the earth's magnetic field

$$= \sqrt{H^2 + V^2}$$

$$= \frac{\mu_o N_p N_s A_s I}{NA} \frac{1}{\theta} \sqrt{\theta_H^{\ 2} + \theta_V^{\ 2}} \qquad ...(5)$$

Measurement of the Angle of Dip

The above two experiments are performed and the angle of dip is given by

$$\tan \delta = \frac{V}{H}$$

Dividing equation (2) by (3), we get,

$$\tan \delta = \frac{V}{H} = \frac{\theta_V}{\theta_H} \qquad ...(6)$$

The calibration of galvanometer is not needed in this case.

Calibration of Ballistic Galvanometer Using Earth Inductor

Equations (2) can be used for the calibration of the ballistic galvanometer, because if H is known we can determine the constant K as given below :

$$\frac{2NAH}{R} = k\,\theta_H \left(1 + \frac{\lambda}{2}\right)$$

$$K = \frac{2NAH}{R\theta_H \left(1 + \frac{\lambda}{2}\right)} \qquad ...(7)$$

MAXWELL'S EQUATIONS

The complete equations of electromagnetic theory are listed below:

$$\nabla . \vec{D} = r \qquad (a)$$

$$\nabla . \vec{B} = 0 \qquad (b)$$

$$\nabla \times \vec{E} = - \nabla . \vec{B} \frac{\delta \vec{B}}{\delta t} \qquad (c) \qquad ...(1)$$

$$\nabla \times \vec{B} = \mu_o \left(\vec{J} + \frac{\delta \vec{D}}{\delta t} \right) \qquad (d)$$

Maxwell equations were not generally accepted for many years after they were postulated (1873). *His curl equations involving* $\nabla \times \vec{E}$ *and* $\nabla \times \vec{B}$ *implied that time-varying electric and magnetic fields it free space (*empty space) *were interdependent, a changing electric field being able to generate a magnetic field, and vice versa.* The radio waves were unknown at the time and it was after 15 years (1888) when Hertz demonstrated that electromagnetic waves were possible as predicted by Maxwell.

Along with Maxwell's equations certain other fundamental relations are of importance in dealing with electromagnetic problems. Among these may be mentioned Ohm's law at a point.

$$\vec{J} = \sigma\vec{E}, \qquad ...(2)$$

The continuity relation

$$\nabla.\vec{J} = -\frac{\delta\rho}{\delta t}, \qquad ...(3)$$

The force relations,

$$\vec{F} = q\vec{E}. \qquad ...(4)$$

$$d\vec{F} = Id\vec{l} \times \vec{B} \qquad ...(5)$$

and Lorentz equation,

$$\vec{F} = q\,(\vec{E} + \vec{v} \times \vec{B}), \qquad ...(6)$$

and the relation between $\vec{E}$ and $\vec{D}$ and between $\vec{B}$ and $\vec{H}$ as given by

$$\vec{D} = \varepsilon\vec{E} = \varepsilon_0\vec{E} + \vec{P}, \qquad ...(7)$$

$$\vec{B} = \mu\vec{H} = \mu_0\,(\vec{H} + \vec{M}) \qquad ...(8)$$

The equations (1) to (8) together with Newton's laws of motion are sufficient to explain any electromagnetic phenomena.

MAXWELL'S EQUATIONS IN FREE SPACE

In the preceding section, Maxwell's equations are stated in their general from. For free space, where the current density $\vec{J}$ and charge density ρ are zero as well as $\vec{D} = \varepsilon_0\vec{E}$ and $\vec{B} = \mu_0\vec{H}$, the equations reduce to a simple differential form,

$$\vec{\nabla}.E = 0 \qquad (a)$$

$$\nabla.\vec{B} = 0 \qquad \text{(b)}$$

$$\nabla \times \vec{E} = -\frac{\delta \vec{B}}{\delta t} \qquad \text{(c)} \qquad ...(1)$$

$$\nabla \times \vec{E} = \mu_o \varepsilon_o \frac{\delta \vec{E}}{\delta t} \qquad \text{(d)}$$

Any electromagnetic field must satisfy all of Maxwell equations. Let us assume that there is a changing charge and current distribution within a certain region of space, and let us consider the fields produced by this source of radiation. In the free space outside the region, the charge and current densities are everywhere zero.

In integral form the equations are

$$\oint_S \vec{E}.d\vec{S} = 0 \qquad \text{(a)}$$

$$\oint \vec{B}.d\vec{S} = 0 \qquad \text{(b)}$$

$$\oint \vec{E}.d\vec{l} = -\int_S \frac{\delta \vec{B}}{\delta t}.d\vec{S} \qquad \text{(c)} \qquad ...(2)$$

$$\oint \vec{B}.d\vec{l} = \mu_o \varepsilon_o \int_S \frac{\delta \vec{E}}{dt}.d\vec{S} \qquad \text{(d)}$$

First two equations are obtained by Gauss's divergence theorem ($\oint \vec{E}.d\vec{S} = \int_v \text{div } Edv$) and last two equations by Stores's theorem ($\oint \vec{B}.d\vec{l}$ $= \int_8 \text{curl } \vec{B}.d\vec{S}$).

ELECTROMAGNETIC WAVES IN FREE SPACE

For free space the charge density $\rho = 0$ and the current density $\vec{J} = 0$ and K (relative permittivity) = 1 and K_m (or μ_r) relative permeability = 1. Therefore, the constitutive equations are,

$$\vec{D} = \varepsilon_0 \vec{E}$$

$$\vec{B} = \mu_0 \vec{H}$$

$$J = 0$$

$$r = 0$$

Maxwell's equations are given by Eq. (1) *i.e.*,

$$\nabla . \vec{E} = 0 \qquad (a)$$

$$\nabla . \vec{B} = 0 \qquad (b)$$

$$\nabla \times \vec{E} = -\frac{\delta \vec{B}}{\delta t} \qquad (c)$$

$$\nabla \times \vec{B} = \epsilon_o \mu_o \frac{\delta \vec{E}}{dt} \qquad (d)$$

To obtain the differential equation of electromagnetic wave propagation we eliminate $\vec{B}$ from equations (1a) and (1b)

Taking the curl of equation (1c), *i.e.*,

$$\nabla \times \nabla \times \vec{E} = -\nabla \times \frac{\partial \vec{B}}{\partial t}$$

$$= -\frac{\delta}{\delta t}(\nabla \times \vec{B})$$

Substituting $\nabla \times \vec{B}$ from eq. (1.20d), we get,

$$\nabla \times \nabla \times \vec{E} = -\frac{\delta}{\delta t}\left(\varepsilon_o \mu_o \frac{\delta \vec{E}}{\delta t}\right)$$

$$= -\mu_o \varepsilon_o \frac{\delta^2 E}{\delta t^2}$$

According to vector product rule,

$$\nabla \times \nabla \times \vec{E} = \nabla (\nabla . \vec{E}) - \nabla^2 \vec{E} = -\nabla^2 \vec{E}$$

Since, $\nabla . \vec{E} = 0$, [by eq. (1(a)] hence above relation reduces,

$$-\nabla^2 \vec{E} = -\mu_o \varepsilon_o \frac{\delta^2 \vec{E}}{\delta t^2}$$

or $$\nabla^2 \vec{E} = \mu_o \varepsilon_o \frac{\delta^2 E}{\delta t^2} \qquad ...(2)$$

Similarly $\vec{E}$ can be eliminated from Eqs. (1c) and (1d) yielding,

$$\nabla^2 \vec{B} = \mu_o \varepsilon_o \frac{\delta^2 E}{\delta t^2} \qquad ...(3)$$

Equations (2) and (3) representing the relation between the space and time variation of $\vec{E}$ and $\vec{B}$ are called "three dimensional wave equation" for $\vec{E}$ and $\vec{B}$ respectively. If we substitute,

$$\nabla^2 = \left(\frac{\delta^2}{\delta x^2} + \frac{\delta^2}{\delta y^2} + \frac{\delta^2}{\delta z^2}\right)$$

and $\vec{E} = \hat{i}E_x + \hat{j}E_y + \hat{k}E_z$

Then equation (1.22) becomes,

$$\left(\frac{\delta^2}{\delta x^2} + \frac{\delta^2}{\delta y^2} + \frac{\delta^2}{\delta z^2}\right)(\hat{i}E_x + \hat{j}E_y + \hat{k}E_z)$$

Similar expression for $\vec{B}$ is obtained if we analyse Eq. (3)

The wave equation given by Eqs. (2) and (3) is sometimes called "D" *Alembert's equation", have been integrated by him.*

Each component of $\vec{E}$ and $\vec{B}$ is identical with the familiar wave equation,

$$\frac{\delta^2 y}{\delta x^2} = \frac{1}{v^2}\frac{\delta^2 y}{\delta t^2}$$

The factor $\mu_o\varepsilon_o$ has the same significance as the $1/v^2$. Dimensionally eq. (2),

$$\frac{\text{volts}}{\text{met}^3} = \mu_o\varepsilon_o \frac{\text{volt}}{\text{met.sec.}^2}$$

since $\quad E = \text{volt/metre}$

or $\quad 1/\sqrt{\mu_o\varepsilon_o} = \text{met/sec.}$

Same result is obtained if we analyse Eq. (3).

Thus it appears that $1/\sqrt{\mu_o\varepsilon_o}$ has the dimensions of velocity and is in fact the velocity with which the wave propagates in the free space. Hence the velocity of wave in free-space $= 1/\sqrt{\mu_o\varepsilon_o}$

$$= \frac{1}{\sqrt{[4\pi\times 10^{-7}\,\text{weber / amp.met.}8.85\times 10^{-22}\,\text{Coul}^2\,/\,\text{Nwt met}^2}}$$

Since 1 Weber $= \dfrac{1\,\text{Newton.metre}}{\text{Amp.}}$

hence $\frac{1}{\sqrt{\mu_o \varepsilon_o}}$

$$= \frac{1}{\sqrt{4\pi \times 10^{-10} \times 8.85 \frac{\text{Newton.metre}}{\text{Amp}^2.\text{met.}} \cdot \frac{\text{Coul}^2}{\text{Newton} - \text{metre}^2}}}$$

$$= \frac{1}{\sqrt{4 \times 3.14 \times 8.85 \times 10^{-19} \frac{1}{(\text{Coul / sec})^2} \cdot \frac{\text{Coul}^2}{\text{net}^2}}}$$

(Since Amp = Coul/Sec.)

$$= \frac{1}{\sqrt{4 \times 3.14 \times 8.85 \times 10^{-19}}} \text{ met./sec.}$$

$$= 3.00 \times 10^8 \text{ met/sec.} \quad ...(4)$$

which is same as the velocity of light in vacuum. Thus we can conclude that the field vectors $\bar{E}$ and B propagate as waves in vacuum (free space) with the velocity of light c. This result suggests that light waves are electromagnetic in nature.

SOLVED EXAMPLES

Example 1:

A copper strip 2.0 cm wide and 1.0 mm thick is placed in a magnetic field of magnitude 1.5 webers/ metre² as z-direction. If a current of 100 amp is set up in the strip along the x-direction, calculate the Hall potential difference across the strip given the number of electrons per cubic metre = 8.4 × 10²⁸ metre³.

Solution:

From equation (1),

$$E_y = \frac{j_x B_z}{Ne}$$

But $E_y = \frac{V}{d}$ where V is the potential difference

and $jx = \frac{i}{A} = \frac{i}{dh}$

where h is thickness.

Substituting these values, we get

$$\frac{V}{d} = \frac{i}{dh} \cdot \frac{B_z}{Ne}$$

$$V = \frac{iB_z}{Neh} = \frac{200 \text{ amp} \times 1.5 \text{ weber / metre}^2}{8.4 \times 10^{28/m^3} \times 1.6 \times 10^{-19} \text{ coul} \times 1.0 \times 10^3 \text{ m}}$$

$$= 2.2 \times 10^{-5} \text{ volt}$$

$$= 22 \text{ micro volt.}$$

Example 2:

A uniform magnetic field of magnitude 1.5 Weber/metre² points horizontally from south to north. A proton of energy 5.0 MeV moves vertically downward through this field. Calculate the force on it.

Solution:

Kinetic energy of the proton is

$$K = 5.0 \text{ MeV} = 5 \times 10^6 \text{ eV}$$

$$= 5 \times 10^6 \text{ eV} \times 1.6 \times 10^{-19} \text{ joule}$$

$$= 8.0 \times 10^{-13} \text{ joule.}$$

But K = 1/2 mv^2, where m is the mass and v is the velocity, or

$$v = \sqrt{\frac{2K}{m}} = \sqrt{\frac{2 \times 8.0 \times 10^{-13}}{1.7 \times 10^{-27}}} = 3.1 \times 10^7 \text{ metres/sec.}$$

Now, according to equation (1.2),

$$F = qvB \sin \theta.$$

As given $\vec{B}$ is horizontal and proton moves vertically downward, hence angle between $\vec{v}$ and $\vec{B}$ is 90°,

therefore, $\sin \theta = \sin 90 = 1$

$F = qvB = (1.6 \times 10^{-19}\text{Coul}) \times (3.1 \times 10^7\text{metre/sec}) \times 1.5\text{weber/}$

metre^2

But, $\text{Weber} = \frac{\text{Newton.metre}}{\text{Ampere}}$

hence, $F = 7.4 \times 0^{-12}$ Netwons.

Since $\vec{F}$ is perpendicular to $\vec{v}$ and $\vec{B}$ from relation,

$$\vec{F} = q\vec{v} \times \vec{B},$$

the direction of this force is east.

Example 3:

A 10 cm long wire carrying a current of 10 amp. is held at an angle 30° with the direction of a uniform magnetic field of strength 1 weber/ metre^2. Calculate the force acting on the wire.

Solution:

According to equation (1), the force is

$$\vec{F} = i\vec{l} \times \vec{B}$$

$$F = ilB \sin \theta.$$

Putting the given values,

$$F = 10 \times 10 \times 10^{-2} \times 1 \sin 30$$

$$= 0.5 \text{ Newton.}$$

Example 4:

If the magnitude of $\vec{H}$ in a plane wave is 1 amp./metre. Find the magnitude of $\vec{E}$ for a plane wave in free space.

Solution:

By equation (1)

$$E = \sqrt{\frac{\mu_0}{\varepsilon_0}} H$$

$$= 376 \times 1 \text{ ohms. Amp/metre}$$

$$= 376 \text{ volts/metre.}$$

Example 5:

When 0.1 coul. of charge is passed through a moving coil B.G. a deflection of 30 mm is observed on a scale one meter away. Find the current sensitivity of the galvanometer if the time period of the coil is 10 seconds.

Solution:

From equation (1)

$$\text{Current sensitivity} = \frac{T}{2\pi} \times \text{charge sensitivity}$$

$$= \frac{10\,\text{sec}}{2\times 3.14} \times \frac{30\text{ mm}}{0.1\text{ Coul}}$$

$$= 477.5\text{ mm/amp.}$$

Example 6:

A condenser of capacity 0.2 μF is charged to 3 volts when discharged through a B.G., it gives a deflection of 10 cm. Calculate the current sensitivity of galvanometer, if its periodic time be 10 sec.

Solution:

$$\text{Charge sensitivity} = \frac{10}{0.2\times 3}\ \frac{\text{cm}}{\text{micro coulomb}}$$

$$= 16.6\ \frac{\text{cm}}{\text{micro coulomb}}$$

$$\text{Current sensitivity} - \frac{2\pi}{T}.\ \text{Charge sensitivity}$$

$$= \frac{2\times 3.14}{10\text{ sec}} \times \frac{16.6\text{ cm}}{\text{micro coul}}$$

$$= 10.4\text{ cm/micro amp.}$$

Example 7:

A circular coil of 100 turns has an effective radius 50 cm and carries a current 0.10 amp. Calculate the work done required to turn it in an external uniform magnetic field 1.5 weber/metre² through 180°.

Solution:

Let us take the coil be turned from position $\theta = 0$ to $\theta = 180°$, the work required is the difference in energy between the two positions, the potential energy $\vec{U} = -\,m.\vec{B}$

Hence, work required $= U_e = 180 - {}^{U}\theta = 0$

$$= -\,mB\text{Cos }180 - (-\,mB\text{ Cos }0)$$

$$= 2mB$$

But $m = NiA$

Hence, work required $= 2\ NiAB$

$$= 2Ni\,(\pi r^2)\,B$$

$$= 2 \times 100 \times (0.10) \times \pi \times (5 \times 10^{-2})\ 1.5$$

$$= 0.24 \text{ joule}$$

Example 8:

A condenser charged to 2 volts is discharged through a ballistic galvanometer, when the corrected deflection is 1.6 cms and the current sensitivity is 4.54 × 10² mm/μA and the periodic time is 12 seconds. Calculate the capacity of the condenser.

Solution:

$$\text{Charge sensitivity} = \frac{T}{2\pi} \times \text{current sensitivity}$$

$$= \frac{12}{2 \times 3.14} \times 4.54 \times 10^2 \text{ mm/micro coul}$$

$$= \frac{12 \times 4.5 \times 10}{2 \times 3.14} \text{ cm/micro coul}$$

But $\quad \text{charge sensitivity} = \dfrac{\text{deflection}}{\text{charge}}$

or $\quad \text{Charge} = \dfrac{\text{deflection}}{\text{charge sensitivity}}$

$$= \frac{9.6 \text{ cm}}{\dfrac{12 \times 4.5 \times 10}{2 \times 3.14} \text{ cm/micro coul}}$$

or $\quad q = \dfrac{9.6 \times 2 \times 3.14}{12 \times 4.5 \times 10}\ \mu \text{ coul}$

$$= \frac{9.6 \times 2 \times 3.14 \times 10^{-6}}{12 \times 4.5 \times 10} \text{ coul}$$

But, $\quad q = c\,V$

or $\quad c = \dfrac{q}{V} = \dfrac{9.6 \times 2 \times 3.14 \times 10^{-6}}{12 \times 4.5 \times 10 \times 2}$ Coul/volt

$$= 20.16 \times 10^{-8} \text{ Farad}$$

$$= 0.2016\ \mu\text{F}.$$

Example 9:

The successive deflections to the right and lift of the mean position in the case of ballistic galvanometer are found to be 25.9 and 24.8 cms. Calculate the logarithmic decrement.

Solution:

As we know that $$\frac{\theta_1}{\theta_2} = \frac{\theta_2}{\theta_3} e^{\lambda}$$

$$e^{\lambda} = \frac{25}{24.9} = \frac{24.9}{24.8} = \frac{25+24.9}{24.9+24.8}$$

$$\lambda = 2.3 \log_{10} \frac{49.9}{49.7} = 0.0039$$

Example 10:

A condenser of 1,000 of is charged to a potential difference of 1 volt and then discharged through a B.G. The first throw an noted on a scale placed 1 meter away is 62.2 cms. If the time period of free vibrations be 10 seconds, and logarithmic decrement 0.02, calculate the ballistic constant of the galvanometer and its figure of merit.

Solution:

$$q = cV$$

$$= 1000 \times 10^{-12} \times 1 = 10^{-9} \text{ coul.}$$

and also $$q = K\theta_1 \left(1 + \frac{\lambda}{2}\right),$$

where, $$K = \frac{T}{2\pi} \frac{c}{NAB}$$

or,

$$10^{-9} = K \times 62.2 \times 10^{-2} \times \left(1 + \frac{0.02}{2}\right)$$

or $$K = \frac{10^{-9}}{62.2 \times 10^{-2} \times \left(\frac{2+0.02}{2}\right)}$$

$$= \frac{10^{-9} \times 1000 \times 100}{622 \times 101} \frac{\text{Coul}}{\text{metre}}$$

Since, $$K = \frac{T}{2\pi}\left(\frac{c}{NAB}\right)$$

In fact, $\frac{c}{NAB}$ is the figure of merit

$$\frac{c}{NAB} = K.\frac{2\pi}{T} = \frac{10^{-9}\times 1000\times 100}{622\times 101}\quad\frac{2\pi}{10}$$

$$= 10^{-10} \text{ amp/radian},$$

Let us consider a deflection of 1 mm on a scale, 100 cm away

Angular deflection $= \frac{10.3}{1} = \frac{1}{1000}$ radian

Current required to produce this deflection,

$$= 10^{-10} \text{ amp/radian} \times \frac{1}{1000} \text{ radian}$$

$$= 10^{-13} \text{ amp.}$$

Example 11:

A 50 ohm coil galvanometer can carry a current 10 MA. How can you convert it into

(a) voltmeter of range 500 volts

(b) ammeter of range 10 amp.

Solution:

(a) Using relation (1), *i.e.*,

$$V = i_g (R_m + R_g) \text{ or } R_m = \frac{V}{i_g} - R_g$$

i_g = 10MA

= 10 × 10–3 amp

R_g = 50 ohms

V = 500 volts

hence, $$R_m = \frac{500}{10\times 10^{-3}} - 50$$

= 49,950 ohms

R_m = 49950, is connected in series.

(b) Using relation (1.30), *i.e.*,

$$i = i_g\left(1+\frac{R_g}{R_s}\right)$$

given,
$$i = 10 \text{ amp}$$
$$i_g = 10 \times 10^{-3} \text{ amp.}$$
$$R_g = 50 \text{ ohms}$$

Hence,
$$10 = 10 \times 10^{-3}\left(1+\frac{50}{R_s}\right)$$

or
$$\frac{10}{10\times10^{-3}} = \frac{R_s+50}{R_s}$$

or
$$1+\frac{50}{R_s} = 1000$$

or
$$\frac{50}{R_s} = 999$$

or
$$R_s = \frac{50}{999} = 0.05005$$

R_s is shunt resistance.

Example 12:

A pointer type galvanometer of 20 ohm resistance has a linear scale of 50 divisions. It gives a deflection of 30 divisions when connected in series with a resistance of 2980 ohms and an accumulator of 2 volts. How will you convert it into an ammeter of range 3 ampere ?

Solution:

The current corresponding to 30 divisions by Ohm's law.

$$i_{30} = \frac{2}{2980+20} = \frac{2}{3000} \text{ amp.}$$

The maximum value of the current galvanometer can carry is corresponding to 50 divisions. Hence,

$$i_g = i_{50} = \frac{2}{3000}\times\frac{50}{30} \text{ amp.}$$

Now using the relation,

$$i = i_g\left(1+\frac{R_g}{R_s}\right)$$

$$3 = \frac{2}{3000} \times \frac{50}{30}\left(1+\frac{20}{R_s}\right)$$

or
$$\frac{3\times3000\times30}{2\times50} = 1+\frac{20}{R_s}$$

or
$$\frac{20}{R_s} = \frac{3\times3000\times30}{2\times50} - 1$$

$$= 2700 - 1 = 2699$$

or

$$R_s = \frac{29}{2699}\ 0.00741 \text{ ohm (shunt)}$$

Example 13:

Paraffin has relative permittivity K = 2.1. Find the index of refraction for paraffin and also the velocity of the wave in paraffin.

Solution:

As paraffin is a non-ferrous material, hence by eq. (1)

$$\eta = \sqrt{K} = \sqrt{2.1} = 1.45$$

By Eq. (1.30) $v = C/\eta = C/\sqrt{K}$

$$= C/\sqrt{2.1} = 2.07 \times 10^6 \text{ metre/sec.}$$

Example 14:

The relative permittivity of distilled water is 81. Calculate refractive index and velocity.

Solution:

$$\eta = \sqrt{K} = \sqrt{81} = 9$$

$$v = C/\sqrt{81} = 0.11\ C$$

$$= 3.33 \times 107 \text{ metres/sec.}$$

EXERCISES

1. Write the Maxwell's equations. Explain the physical significance of each equation.

2. Explain different types of currents you know. Explain fully the concept of displacement current.
3. What are energy losses in a generator? How these are reduced?
4. Describe the principal and construction of an D.C. electric motor. Define the efficiency of an electric motor. Prove that for maximum output, the back emf should be half the applied voltage.
5. Describe with necessary theory the use of an earth inductor to find the angle of dip.
6. Describe with relevant theory the method for measuring strong magnetic field.
7. Write Maxwell's equations. Solve them in free space.
8. Describe the principle, construction and working of an A.C. generator or dynamo.
9. What do you mean by a D.C. generator? Describe its working.

4

Magnetic Deflection Sensitivity

INTRODUCTION

The magnetic deflection sensitivity is defined as deflection in metres per weber square metre.

The magnetic field deflects the beam perpendicular to is direction of motion and the electron follows the circular path in the field. The magnitude of the deflecting force eB_v from relations $\vec{F} = e(\vec{v} \times \vec{B})$ where B is the magnitude of the magnetic field and v is the velocity of electron, say radius of the circular path is R given by centrifugal force.

$$\frac{mv^2}{R} = eBv$$

$$R = \frac{mv}{eB} \qquad ...(1)$$

Let the magnetic field B be uniform in a region of length l in the direction of beam and zero outside this region. The electron after leaving the field travels in a straight line tangential to the circular path.

If q is the angular deflection suffered by beam due to the magnetic field and considering that q is small

$$\theta = \frac{l}{R} = \frac{Y}{(L + l/2)} \qquad ...(2)$$

where L is the distance from edge of l to the screen. By eqn. (1),

$$\frac{Y}{(L + l/2)} = \frac{l}{R} = \frac{leB}{mv} \qquad ...(3)$$

$$\text{Deflection sensitivity} = \frac{Y}{B} = \frac{el\,(L + l/2)}{mv}$$

If electron is accelerated to the velocity v by accelerating potential V_a

$$\frac{1}{2} mv^2 = eV_a$$

$$m^2v^2 = 2meV_a$$

$$mv = \sqrt{2mev_a}$$

Hence deflection sensitivity

$$\frac{Y}{B} = \frac{el(L+l/2)}{\sqrt{2me\ V_a}}$$

or
$$\frac{Y}{B} = (L+l/2)\sqrt{\frac{e}{m}}\ \frac{l}{\sqrt{2V_a}} \quad ...(4)$$

Thus deflection sensitivity is proportional to l, L + l/2 and Ö e/m inversely proportional to the square root of the final anode voltage.

The comparison of equations (1) and (4) shows that the effect of changed the accelerating potential V_a for changing the brightness of the spot will be less, when magnetic field is used for deflection than when electric field is used. *So the brightness of the spot can be increased with smaller decrease insensitivity for magnetic C.R.O. than for electric C.R.O.*

TIME BASE

In many applications of the C.R.O., the movement of the spot in one direction generally the horizontal is required to be proportional to time. The arrangement used to achieve this is called time base. By using time base the base line of any pattern on the screen can be regarded as time axis and the pattern will record the variation of the quantity under study with time.

Time Base Circuit

A neon lamp is connected to the terminals of condenser c. The condenser is charged through R by D.C mains and discharged through neon lamp. The condenser is charged until the voltage across it reaches the critical value needed to glow the neon lamp. When the neon lamp glows, it short circuits the condenser which thus discharges through the lamp causing the rapid fall of the potential and the spot on screen returns to the initial position suddenly. The process is repeated, the new cycle starts again, and the spot traces the same line several times per second and looks like a continuous line. The neon lamp will flash at regular interval depending upon the product RC. The frequency of time base can be calculated assuming that the charging is linear. If V is the voltage of the D.C. mains,

$$i = V/R$$

charge on the condenser Q = it, where t is tie of the charging. The voltage on the condenser V' which is the same as the striking voltage for the neon lamp

$$V' = Q/C$$

But $$t = \frac{Q}{i} = \frac{V'CR}{V}$$

and the frequency $$f = \frac{1}{t} = \frac{V}{V'CR}$$

f can be made to vary by changing C and R within certain limits.

USES OF CATHODE RAY OSCILLOSCOPE

Cathode ray oscilloscope is widely used in number of measurements and few of them are given below:

1. *In television.* In television receiver C.R.O. is used for recreation of images.
2. *In radar.* IN radar C.R.O. is used to detect the position of the target such as aircraft and then to give the visual indication.
3. *In industry.* C.R.O. is used for study of mechanical pressures and indicator diagrams of internal combustion engines and moisture contents in a factory
4. *In medicine and surgery.* C.R.O. is used for electrocardiography *i.e.*, the study of the hears's action.
5. *In engineering.* It is used to detect faults in generators, motors and in locating the faults in radio receivers, transmitters and other radio equipment.
6. *Measurement of Direct or Alternating Voltage.* As we know, the deflection on C.R.O. screen is directly proportional to the voltage applied to the deflecting plate. First the deflection sensitivity is calculated *i.e.*, the deflection on the screen is directly calibrated in terms of known voltage. The direct voltage to be measured

is applied between a pair of deflecting plates. Measure the deflection produced on the screen when this deflect in is multiplied by the deflection sensitivity gives the values of direct voltage.

The alternating voltage to be measured is applied between a pair of deflecting plates and the length of the straight line produced on the screen is measured knowing the deflection sensitivity, peak to peak value of the applied voltage can be calculated.

7. *Measurement of Direct or Alternating Current.* As we know that C.R.O. is a device working on the voltage applied to the deflecting plates, therefore, the current to be measured is first passed through a known resistance and then voltage developed is applied to one pair of the deflection plates. The voltage is calculated as in the pervious case 6 from the known deflection on the screen. The current can easily be calculated by Ohm's law.
8. *Measurement of Frequency.* The unknown frequency can be determined by comparison with a known frequency. Suppose we want to determine the frequency of an alternating current, this unknown frequency (current) is applied to one pair of deflecting plates and a known frequency from some variable frequency, standard oscillator is applied to the other deflecting plates. The frequency of standard oscillator is varied until a single loop stationary pattern is obtained. Then the frequency of sinusoidal current is the same as that of the oscillator.

CHARGED PARTICLE IN UNIFORM (STATIC) ELECTRIC FIELD APPLIED ALONG THE DIRECTION OF THE PARTICLE MOTION

Let a particle of charge q be placed in an electric field $\vec{E}$. Since $\vec{E}$ is the force per unit charge (newtons per unit coulomb,), the force $\vec{F}$ on the particle is given by,

$$\vec{F} = q\,\vec{E} \qquad ...(1)$$

The force is in the same direction as the field $\vec{E}$ if the charge is positive and opposite to the field if the charge is negative.

When the field is applied, the particle is accelerated uniformly in the direction of field. According to Newton's second law the force on a particle of mass m is given by,

$$\vec{F} = m\vec{a} = q\vec{E}$$

where a is acceleration and given by,

$$\vec{a} = \frac{q}{m}\vec{E} \quad ...(2)$$

$$\frac{d^2\vec{r}}{dt^2} = \frac{q}{m}\vec{E} \quad ...(3)$$

As $\vec{E}$ is constant on integrating eqn. (3) we will get velocity of the particle,

$$\frac{d\vec{r}}{dt} = \vec{v} = \frac{q\vec{E}}{m}t + C \quad ...(4)$$

where C is the constant of integration when t = 0, $\vec{v} = \vec{u}$ (initial velocity of the particle).

From equation (4), C = u

Substituting C in equation (4), it reduces,

$$\vec{v} = \frac{q\vec{E}}{m}t + \vec{u} \quad ...(5)$$

Again integrating (5), we have,

$$\vec{r} = \frac{q\vec{E}}{2m}t^2 + \vec{u}t + C_1 \quad ...(6)$$

when t = 0. Suppose the position of particle is given by $\vec{r}_o$, $C_1 = r_o$.

Substituting C1 in equation (6), we get,

$$\vec{r} = \frac{q\vec{E}}{2m}t^2 + \vec{u}t + \vec{r}_o \quad ...(7)$$

The relation (7) is applicable in general whatever may be direction of $\vec{r}_o$, $\vec{u}$ and $\vec{E}$. $\vec{AD}$ is the displacement $\vec{r}$ at any time t.

If the charge is placed at rest on the origin of the coordinates, then,

$$\vec{u} = 0, r_o = 0$$

We have from equation (7),

$$\vec{r} = \frac{q\vec{E}}{2m}t^2 \quad ...(8)$$

and from equation (5),

$$\vec{v} = \frac{q\vec{E}}{2m}t \quad ...(9)$$

Energy Acquired

The field imparts energy to the charged particle. The force is given by,

$$F = m\vec{a} = q\vec{E}$$

The work done by this force in moving the particle gives the energy W acquired by the particle. Thus,

$$W = m\int_1^2 \vec{a}.d\vec{l} \quad (\text{since } w\ \vec{F}.d\vec{l})$$

Substituting $\vec{a} = \frac{d\vec{v}}{dt}$

and $d\vec{l} = \vec{v}.dt$

$$W = m\int_1^2 \frac{d\vec{v}}{dt}.\vec{v}.dt = m\int_1^2 \vec{v}.d\vec{v}$$

If V is the potential difference between the points, then,

$$W = qV$$

Thus,

$$W = \frac{1}{2}m(v_2^2 - v_1^2) = qV \quad ...(10)$$

where v_1 and v_2 are the initial and final velocities of the particle.

If the particle starts from rest, the initial velocity is zero ($v_1 = 0$) so,

$$W = qv = \frac{1}{2}mv^2 \quad ...(11)$$

where v = final velocity. Solving for v

$$v = \sqrt{\frac{2qv}{m}} \text{ metre / sec.} \quad ...(12)$$

on squaring equation (9), we have,

$$v^2 = \frac{q^2E^2}{m^2}t^2$$

$$\text{Kinetic energy} = \frac{1}{2}mv^2 = \frac{1}{2}m.\frac{q^2E^2}{m^2}t^2$$

But from equation (8),

$$\frac{qE}{2m}t^2 = r$$

Hence, Kinetic energy = qEr ...(13)

Electron Volt

If q = e, *i.e.*, particle is electron, then the energy acquired by an electron,

$$W = eV$$

Further, if V = 1 volt,

$$W = e = 1.6 \times 10^{-19} \text{ joule}$$

This amount of energy is called one electron-volt.

$$1 \text{ eV} = 1.6 \times 10^{-19} \text{ joule} \quad ...(14)$$

This is a convenient unit in designating the energies of particles.

MOTION OF A CHARGED PARTICLE IN UNIFORM TRANSVERSE ELECTRIC FIELD

In this case we shall examine the motion of charged particles when the uniform electric field is perpendicular to the direction of its initial velocity.

A beam of charged particles (*e.g.*, electrons beam) moving along x-axis through a pair of parallel plates of length l and separated by distance d. If V be the potential difference between the plates, then E_y = V/d. The force on the charged particle = qE^y.

The force is along the direction of the field *i.e.*, Y-axis, hence the initial velocity of the particle v_x is not affected by this force. So v_x remains constant. If the electron travels a distance x along x-axis in the field, the time t required to cover this distance is given by,

$$t = \frac{X}{v_x} \quad ...(1)$$

In this time, the transverse velocity gained by the particle is,

$$\frac{dy}{dt} = v_y = 0 + at$$

$$= \frac{qE_y}{m}\frac{x}{v_x} \quad ...(2)$$

where $a = \frac{qE_y}{m}$ is acceleration of the particle and initial velocity along x-axis is zero.

The displacement along y – axis is,

$$y = \frac{1}{2}at^2 = \frac{1}{2}\frac{qE_y}{m}\left(\frac{x}{v_x}\right)^2$$

$$= \frac{qE_y}{2mv_x^2}x^2 \qquad ...(3)$$

The quantity $\frac{qE_y}{2mv_x^2}$ is constant, say A, then

$$y = Ax^2 \qquad ...(4)$$

Which represents a parabola. Thus the path of a charged particle in a transverse uniform electric field is parabola. The direction of motion is changed due to field. At any instant t if the inclination of the direction of motion to x-axis be f, then from equation (3),

$$\tan\phi = \frac{dy}{dx} = \frac{qE_y}{2mv_x^2}2x$$

$$= \frac{qE_y}{mv_x^2}x \qquad ...(5)$$

But the particle remains in the field for a distance l. The length of the plates. Hence, the displacement (y_l) along the y-axis due to entire field will be obtained by substituting x = l in equation (3),

$$y_l = \frac{qE_y}{2mv^2}l^2 \qquad ...(6)$$

After traversing the field, the direction of the particle is changed through the angle θ.

$$\tan\theta = \frac{qE_y}{2mv_x^2}l \qquad ...(7)$$

(by putting $\phi = \theta$, x = l in equation (5)

It is to be noted that as soon as the particle emerges from the field, no force acts on it. The particle travels (neglecting gravity) toward screen in a straight line tangent to the parabola at the exit point. If screen is placed at a distance L from the end A of the plate, then the displacement y_2 along y-axis is given by,

$$y_2 = L \tan \theta \quad ...(8)$$

Now, the total displacement on the screen is,

$$y = y_1 + y_2 \text{ [Substituting from eqn. (6) and (8)]}$$

$$y = \frac{qE_y l^2}{2mv_x^2} + L \tan \theta$$

Substituting tan θ from equation (7),

$$y = \frac{qE_y l^2}{2m\, v_x^2} + L\,\frac{qE_y}{m\, v_x^2} l \quad ...(9)$$

Such a deflected electron beam when falls on the screen makes a "visible small luminous spot". This is the principle of the electrostatics cathode ray oscilloscope" which we shall discuss later on.

CHARGED PARTICLE IN A CONSTANT UNIFORM MAGNETIC FIELD

When a charged particle moves through a magnetic field the Lorentz force acts on it. If the charge q is moving with velocity v in a magnetic field $\vec{B}$, the force acting on the particle is given by,

$$\vec{F} = q(\vec{v} \times \vec{B}) \quad ...(1)$$

From Newton's second law, F = ma. Thus,

$$m\vec{a} = q(\vec{v} \times \vec{B}) \quad ...(2)$$

where $\vec{a}$ is acceleration and given by,

$$\vec{a} = \frac{q(\vec{v} \times \vec{B})}{m} \quad ...(3)$$

According to eqn. (3), the acceleration is normal to the plane containing the particle path (*i.e.*, the velocity $\vec{v}$) and $\vec{B}$. We can draw the following important conclusions:

(1) From eq. (3), the acceleration is maximum when angle between $\vec{v}$ and $\vec{B}$ is 90° *i.e.*, if the direction of particle path (indicated by v) is normal to $\vec{B}$.

(2) If $\vec{v} = 0$, *i.e.*, if the particle is at rest, the acceleration $\vec{a}$ (as well as force $\vec{F}$) is zero. It means field has no effect.

(3) If angle between $\vec{v}$ and $\vec{B}$ is zero, then acceleration $\vec{a} = 0$. Thus, if the particle is moving in the same direction as $\vec{B}$, field $\vec{B}$ has no effect as $\vec{a}$ is zero. It means the particle moves with the same velocity *i.e.*, $\vec{v}$.

Thus, the field $\vec{B}$ affects the motion of the charged particle only when the path or the velocity $\vec{v}$ has a component normal to $\vec{B}$. Further if a magnetic field of large extent is at right angles to the direction of motion of a charged particle, the particle is deflected into a circular path. If, however, the velocity v of the particle be inclined to the magnetic field at an angle other than 90° *i.e.*, not normal to the field, it can be resolved into two components. $\vec{v}_1$ parallel to the field, $\vec{B}$and $\vec{v}_2$ perpendicular to $\vec{B}$ and v_2 is normally outward from the page. According to the cross product of $\vec{v}_2 \times \vec{B}$, the acceleration $\vec{a}$ is upward so that positively charged particle describes a curved path in the clockwise direction in the plane of the page.

The magnitude of the force on the particle

$$= qv_2B$$

If particle moves in a curved path of radius R, this force is balanced by centrifugal force,

$$\frac{mv_2^2}{m} = qv_2B$$

$$R = \frac{mv_2}{qB} \qquad ...(4)$$

It is the radius of the curvature of the path of the particle moving in the magnetic field. This is called gyro radius or cyclotron radius. Thus, the larger the velocity of the particle or the larger its mass, the greater the radius. On the other hand, the larger the charge or $\vec{B}$, the smaller the radius.

Since the force acts perpendicular to v_2 hence it changes its direction but not magnitude. And, as the component v_1 of the velocity, parallel to the field $\vec{B}$ remains constant, so that particle moves in the direction of $\vec{B}$. Therefore the combined effect of the circular and linear horizontal motion is to produce a helical motion, with the axis of the helix parallel to $\vec{B}$

Suppose the initial velocity is perpendicular to the magnetic field *i.e.*, $v_1 = 0$, in other words particle is moving in a direction perpendicular to $\bar{B}$. Then the curved path is exactly circular.

The angular velocity,

$$\omega = \frac{v_2}{R}$$

$$\omega = \frac{qB}{m} \text{(by eq. 4)} \qquad ...(5)$$

The time required to complete one circle is period is revolution, is given by,

$$T = \frac{2\pi}{\omega} = \frac{2\pi m}{qB} \qquad ...(6)$$

The number of revolutions per second by a charged particle in a magnetic field is termed gyro frequency or cyclotron frequency is,

$$n \frac{1}{T} = \frac{qB}{2\pi m} \text{ r.p.s.} \qquad ...(7)$$

This relation suggests that cyclotron frequency of charged particle is independent of its velocity and hence also of the energy of the particle so long as the velocity lies in the non-relativistic region *i.e.*, so long as $v < C$. But if particle velocity is comparable to the velocity of the light, we cannot overlook the fact, that mass changes with velocity according to the relation,

$$m \frac{m_o}{(1 - v^2 / c^2)^{1/2}}$$

Pitch of the Helix

It is defined as the distance travelled by the charged particle along $\vec{B}$ in the time in which it completes one revolution, *i.e.*,

Pitch of helix $= v_1 T$

$$= \frac{2\pi m v_1}{qB} \qquad ...(8)$$

DEFLECTION SENSITIVITY IN ELECTRIC FIELD

The electric deflection sensitivity is defined as the amount of deflection of the electron spot produced when a voltage of one volt d.c. is applied between the corresponding deflection plate. This deflection system, l is the length of deflecting plates and L is the distance from edge of the plate to the screen, then the distance of the mid-point O of the field from the screen is L + l/2 and displacement on screen is Y.

$$\text{Tan}\ \theta = \frac{Y}{(L + l/2)}$$

by equation (1) *i.e.*,

$$\text{Tan}\ \theta = \frac{qE_y l}{mv_x^2}$$

Here charge on electron is e an say deflecting field is E_d and electron enters the deflecting plates with velocity v_x

$$\text{Tan}\ \theta = \frac{Y}{L + l/2} = \frac{eE_d l}{mv_x^2}$$

If the potential difference between the plates is V_d and these are separated by distance d then

$$E_d = \frac{V_d}{d}$$

therefore, $$\frac{Y}{L + l/2} = \frac{eV_d l}{dmv_x^2}$$

deflection sensitivity $$= \frac{\text{deflection}}{\text{deflecting voltage}}$$

$$\frac{Y}{V_d} = \frac{e(L + l/2)l}{mv_x^2 d}$$

If the electron beam before entering the deflecting plates was accelerated to achieve this velocity v_x by accelerating voltage V_a

$1/2\ mv_x^2 = eV_a$, substituting this we get

$$\frac{Y}{V_d} = \frac{e(L + l/2)l}{2eV_a d}$$

deflection sensitivity $$= \frac{l}{2d V_a}(L + l/2) \quad ...(2)$$

Thus the deflection sensitivity is

(1) directly proportional to the length of the deflection plates,

(2) directly proportional to (L + l/2)

(3) inversely proportional to the final accelerating **voltage**

(4) inversely proportional to the spacing d between the plates.

The sensitivity proportional to the accelerating voltage V_a. This implies that by increasing the v_x which is done for increasing the brightness of the spot, the deflection sensitivity decreases.

CHARGED PARTICLE IN PARALLEL MAGNETIC AND ELECTRIC FIELDS

Suppose the magnetic field $\vec{B}$and electric field $\vec{E}$are along Y. axis and the particle of charge q is moving along z-axis with velocity v, therefore,

$$\vec{B} = B\hat{j},\ \vec{E} = E\hat{j} \text{ and } \vec{v} = -v\hat{k}$$

The force due to magnetic field on charged particle is directed in a direction perpendicular to B and v *i.e.*, along x-axis,

the magnitude = qBv

If a' is acceleration then equation of motion is

$$ma' = \frac{md^2x}{dt^2} = qBv$$

or

$$a' = \frac{d^2x}{dt^2} = \frac{qBv}{m} \qquad ...(1)$$

The acceleration may be assumed to be constant if the change in magnitude and direction of the velocity due to the two fields is small. Initially particle did not have any velocity along the x-axis *i.e.*, $\frac{dx}{dt} = 0$. Hence, velocity attained after time t is given by

$$v_x = \frac{dx}{dt} = a't \qquad ...(2)$$

and distance travelled in time t, if represented by

$$x = \frac{1}{2}a't^2 \qquad ...(3)$$

If the magnetic field is applied in a region of length l', then the time of travel in the magnetic field is t = l'/v.

Therefore, the velocity of the particle after the emergence from the magnetic field by Eq. (2) is

$$v_x = a'l'/v$$

and its displacement x_1 by eq. (3),

$$x_1 = \frac{1}{2} a' \left(\frac{l'}{v}\right)^2 \quad ...(4)$$

Now putting the value of a' from (2) in (4),

$$x_1 = \frac{1}{2} \frac{qB}{mv} l'^2 \quad ..(5)$$

If the screen in placed at a distance L' from the point where the magnetic field ends. The time required by the electron to reach the screen is L'/v.

So the electron is further displaced through

$$x_2 = v_x L'/v$$

$$= a' \frac{l'}{v} . \frac{L'}{v}$$

$$= \frac{qBv}{m} . \frac{l'}{v} . \frac{L'}{v}$$

$$= \frac{qBv}{m} l'L'$$

Total displacement on the screen

$$x = x_1 + x_2$$

$$x = \frac{qB}{mv}\left(\frac{l'^2}{2} + l'L'\right) \quad ...(6)$$

The electric field is along the Y-axis, acceleration is given by

$$a = \frac{d^2y}{dt^2} = \frac{qE}{m}$$

(force on a charge q due to field intensity is qE)

We have solved it in article (2.2). And the total displacement is given, *i.e.*,

$$y = y_1 + y_2 = \frac{qE}{mv^2}\left(\frac{l^2}{2} + lL\right) \quad ...(7)$$

where l is the length of the electric field region and L the distance of the screen from the end of the electric field eliminating v from eqns. (6) and (7),

$$\frac{x^2}{y} = \frac{q}{m} \cdot \frac{B^2}{E} \frac{\left(l'^2/2 + l'L\right)^2}{\left(l^2/2 + lL\right)}$$

$$= A$$

$$x^2 = Ay \qquad ...(8)$$

This represent a parabola. The particles of same q/m but of different velocities will lie on different points of a parabola. Thomson used these parabolas to determine e/m for positive ions.

Now the ratio y/x

$$\frac{y}{x} = \frac{E}{vB}\left[\frac{\dfrac{l^2}{2} + lL}{\dfrac{l'^2}{2} + l'L}\right]$$

and q/m is eliminated.

$$y = mx$$

This equation represents a straight line. Therefore all the particles with came velocity fall on a straight line passing through the centre of the screen regardless of the different values of q/m.

POSITIVE RAYS (OR CANAL RAYS) AND CATHODE RAYS

The detailed analysis of the phenomena in the discharge tube at low pressure has brought to light the existence of two important radiations called the cathode and positive rays. Under the influence of the electric field some of the gas molecules in the tube will be ionised, forming positive and negative ions. The process of ionisation expelled the electrons from the neutral molecules. The molecules which have been deprived of their electrons will become positively charged and constitute the positive ions. The electrons are repelled away from the cathode and positive ions accumulated at near the cathode which produces the intense field gradient and large potential gradient there.

The accelerated positive ions strike the cathode and cause it to liberate a great number of elections. These electrons travel some distance accelerated through the strong electrical field near the cathode before they acquire sufficient energy to ionize gas molecules by collision. As moving electrons do not emit light by themselves the space, they travel before colliding with molecules will be dark, and this is called the

"Crookes dark space." When the pressure is reduced to the order of 0.01 mm the Crookes dark space practically fills the tube producing fluorescence on the walls of the tube. These results appear to be produced by some radiations emitted from the cathode surface and are called "cathode rays". If the cathode of discharge tube is perforated, some rays also pass through the cathode to the opposite side. These are called positive rays (canal rays) because they travel from anode towards cathode *i.e.*, opposite to cathode rays. The positive ions are attracted by the cathode and pass through it with velocity forming the positive rays. The positive rays are deflected in electric and magnetic field but the deflection is being opposite to that of cathode rays.

DETERMINATION OF E/M OF POSITIVE RAYS BY J.J. THOMSON METHOD (THOMSON MASS SPECTROGRAPH)

Theory

It consists essentially in subjecting a beam of positive ions simultaneously to parallel electric and magnetic fields of uniform intensities and registering the resultant effect on a photographic plate, *i.e.*, one has to examine the motion of positive ions in parallel E and B. We have discussed in the previous article and obtain a relation between the deflection produced in two fields for a charge q of mass m, here we shall assume that charge on ion is e.

$$\frac{x^2}{y} = \frac{e}{m}\frac{B^2}{E}\frac{\left(\frac{l'^2}{2} + l'L\right)^2}{\left(\frac{l^2}{2} + lL\right)} \qquad ...(1)$$

From Eq. (1) replacing q by charge of ion e.

If the length of the magnetic field region is equal to the length of the electric field region, we have

$$l' = l$$

Eq. (1) reduces,
$$\frac{x^2}{y} = \frac{e}{m}\frac{B^2}{m}l\left(\frac{1}{2} + L\right)$$

or
$$\frac{x^2}{y} \alpha \frac{e}{m}$$

which represent a parabola when e/m is constant. These parabolae obtained on the photographic plate. From the knowledge of these parabola traces and constants of apparatus e/m can be calculated.

COMPARISON OF E/M BY ANALYSIS OF TRACES.

Let the parabola corresponding to ions of mass m_1 be pp' while the ion of mass m_2 but of the same charge be qq", the distances pp' and qq' are corresponding to same value y (by cutting the parabola for the same value of y).

$$\frac{\dfrac{e}{m_1}}{\dfrac{e}{m_2}} = \frac{x_1^2 / y}{x_2^2 / y}$$

$$\frac{m_2}{m_1} = \frac{x_1^2}{x_2^2} = \frac{(2x_1)^2}{(2x_2)^2} = \frac{(pp')^2}{(qq')^2}$$

$$m_2 = m_1 = \frac{(pp')^2}{(qq')^2} \qquad ...(1)$$

m_2 can be calculated in terms of m_1. Eq. (1) represents that masses of the ions are inversely proportional to the square of the distance from the axis of the parabola. If there are number of isotopes present, they will be deposited into various parobolae. The distance from the axis parabola is greater for ion of larger mass.

Limitation of the Parabola Method

It is not an accurate method because of the following reasons:

1. Since positive ions are lost by collisions while passing through the narrow tube in the cathode, so the intensity available for photography is very much reduced. This intensity is further reduced as particles spread out into a parabolic are.
2. The traces on the photographic plate are blurred with no definite edges and hence not suited for accurate measurements because the wider slits are used to increase the intensity of parabolic are.
3. The influence of the secondary rays make analysis difficult. Though the pressure inside the tube is as low as possible, is never entirely negligible, the positive ions under test may make collisions

and so gain or loss electrons while passing through the deflecting field. This results in "secondary rays" which may be of different types. Some appear on the plate as a general fog others as straight beams, others even as parabolae which may be mistaken for the genuine once.

4. Resolving power is much less than other mass spectrographs discussed below.

ASTON'S MASS SPECTROGRAPH

It is an improvement on Thomson method to increase the accuracy by modifying the principle of operation.

Principle

The main modification introduced is that the electric and magnetic fields are not applied simultaneously and in parallel direction as Thomson did, but successively, first the electric field and then the magnetic field. Their directions being at right angles to each other. Therefore, the electric field produces a dispersion of the positive rays with respect to the velocity, while the magnetic field brings to a focus the dispersed rays.

Apparatus and its Working

A narrow beam of positive ions is made to pass through two parallel narrow slits A and B (0.01 mm in radius and 20 cm. apart) and a very fine beam is obtained. This beam is subjected to a uniform electric field intensity maintained between the plates C and D. The beam as a whole is deflected towards the negative electrode D. The beam is also dispersed through a certain angle because the beam contains ions of different velocities. The fast moving ions remain for a shorter time under the influence of electric field will be deflected less than the slower ones taking a longer time to traverse the field. Thus the beam is spread out into an electric spectrum according to different velocities. The divergent beam limited by slit S next enters the magnetic field $B \perp E$ and the beam is bent in a direction opposite to that caused by E. Again the slowly moving ions are deflected more than the faster ions. Thus the beam can be focused at some point on photographic plate. If S is the linear displacement of the ions and l the length of their path in the electric field, then from eq. (2),

$$S = \frac{eEl^2}{2mv^2}$$

(Force eE = ma, a = eE/m, initial velocity is zero

$$S = \frac{1}{2}at^2, \; t = \frac{l}{v}, S = \frac{1}{2}\frac{eE}{m}\frac{l^2}{v^2}\Bigg)$$

Similarly if S' is linear displacement and l' the length of path of the ion in the magnetic field

$$S' = \frac{1}{2}\frac{eB}{mv}l'^2$$

$$\Bigg(\text{Force} = evB = ma', \; a' = \frac{evB}{m}, S' = \frac{1}{2}a't'^2, t' = \frac{l'}{v}$$

$$S' = \frac{1}{2}\frac{evB}{m}\frac{l'^2}{v^2} = \frac{1}{2}\frac{eB}{mv}l'^2\Bigg)$$

Let θ and ϕ be the mean angles of deviation in the electric and magnetic fields respectively,

$$\theta = S/l \quad \text{and} \quad \phi = S'/l'$$

$$\theta = \frac{eEl}{2mv^2} \qquad \phi = \frac{eBl'}{2mv}$$

or
$$\theta = K_e \frac{e}{mv^2} \qquad ...(3)$$

and
$$\phi = K_m \frac{e}{mv} \qquad ...(4)$$

where K_e and K_m are constants, depending on the strength and distribution of electric and magnetic fields.

Let $d\theta$ be the dispersion angle due to electric field and $d\phi$ the convergence angle due to the magnetic field. Differentiating equations (3) and (4), we get,

$$d\theta = -\; 2K_e \frac{e}{mv^3}\,dv = -\,2\theta\frac{dv}{v}$$

or
$$\frac{d\theta}{\theta} = -2\frac{dv}{v} \qquad ...(5)$$

Similarly,

$$d\phi = -\,K_m \frac{e}{mv^2}dv$$

$$= -\phi\frac{dv}{v}$$

or $$\frac{d\phi}{\phi} = -\frac{dv}{v} \qquad ...(6)$$

Comparing Eqs. (5) and (6), we get

$$\frac{d\theta}{\theta} = 2\frac{d\phi}{\phi} \qquad ...(7)$$

Let OR = a, and RL = b

The dispersion of the selected group at R is $ad\theta$. In the absence of magnetic field the dispersion produced in the beam for a distance a + b

$$= (a + b)\ d\theta$$

As this dispersion (or divergence) is annulled by the magnetic field acting at R so that the group comes to focus at L, distance b from R the condition of focusing is given by,

$$(a + b)\ d\theta = bd\phi$$

$$\frac{a+b}{b} = \frac{d\phi}{d\theta} \qquad ...(8)$$

$$\frac{a+b}{b} = \frac{\phi}{2\theta}, \text{ from eqn. (7)}$$

$$(a + b)\ 2\theta = b\phi$$

$$b\ (\phi - 2\theta) = 2a\theta \qquad ...(9)$$

This is the condition for focusing. If plate P is inclined at an angle α to the incident beam AB and RN be the perpendicular on the line joining O and plate P, from triangle RNO

$$RN = a \sin (\alpha + \theta) \qquad ...(10)$$

and from triangle RNL,

$$RN = b \sin RLN$$

$$RN = b \sin (\phi - \alpha - \theta) \qquad ...(11)$$

Equating eqns. (10) and (11),

$$a \sin (\alpha + \theta) = b \sin (\phi - \alpha - \theta)$$

for small angle,

$$a(\alpha + \theta) = b(\phi - \alpha - \theta) \qquad ...(12)$$

Comparison of eqns. (9) and (12) shows that two equations are identical if $\alpha = \theta$. *Thus the condition for focussing is that photographic plate must be placed at an angle θ with the direction of incident positive ray.*

Separation of Isotopes by Analysis of Traces

Substance of known atomic masses are mixed up with elements whose isotopes are to be determined and traces are obtained on photographic plate. The traces of the known atomic mass serve as reference, thus the traces of unknown isotopes are known by measuring the distance from the reference point.

Advantages of Aston's Mass Spectrograph Over Thomson Spectrograph

The modified technique of Aston have the following advantages over Thomson method:

1. The dispersion of the ionic beam according to different velocities of ions is greater than in Thomson mass spectrograph.
2. All the ions having the same e.m are focussed at one pint, instead of spreading them into a parabola thus securing greater intensity for photography. Hence a fine slit can be used to obtain sharp lines to be obtained on the photographic plate, thereby securing greater sensitivity. The accuracy was 1 in 1,000.

180° MAGNETIC FOCUSSING

Let a beam of charged particles enters a region through a slit in which there be a uniform magnetic field $\vec{B}$ perpendicular to the beam. The particles will be deflected with a radius of curvature R given by the relation

$$\frac{mv^2}{R} = qBv$$

$$R = \frac{mv}{qB}$$

Where v is the component of velocity in the plane of the paper *i.e.*, normal to B. If we examine the beam after 180° of the motion, we find that it is spread out in the plane on motion because the different particles with different masses and velocities describe the circular path of different radii. Thus we can obtain a beam of particles having closely equal momentum after 180° deflection, provided that the particles have the same charge. Therefore, such device can be used as "momentum selector".

One advantage of using a deflection of 180° is that particles of equal momenta, nut moving through an entrance slit at slightly different angles are brought to an approximately common focus after 180°. The particle 2 after describing circular path reaches at Q and SQ = 2R, the diameter of the circle. If the particle is making an angle θ to 2, say 1 or 3 the it will reach at p after describing an are of circle whose diameter is inclined to SQ at an angle θ. The width of the focussed image is,

$$PQ = SQ - SP$$
$$= 2R - 2R \cos \theta$$
$$= 2R (1 - \cos \theta)$$

and $\cos \theta = 1 - \theta^2/2 + \theta^4/4.$

for small value of θ, then θ^2 be very small hence width PQ is small. 180° magnetic focussing is also a method used in mass spectrographs.

DEMPSTER MASS SPECTROGRAPH

Dempster in 1918 devised this spectrograph suitable for analysis of substances available in solid form.

Principle

It is based on 180° magnetic focussing method. Therefore, in this method there is only one deflecting agency, namely the magnetic field, although an electric field is used to accelerate the ions.

Suppose the ions are produced with negligible velocity and these ions are accelerated by the same potential difference and so possessed the same velocity and energy. When a magnetic field $\vec{B}$ is applied at right angle sot the direction of motion of these accelerated ions, the ions will describe a circular path.

Let ions acquire velocity v by accelerating potential V,

$$\frac{1}{2} mv^2 = Ve \qquad ...(1)$$

R is radius of curvature of circular path is given by,

$$mv^2 /R = Bev \qquad ...(2)$$

$$e/m = v/BR$$

$$\frac{e}{m} = \frac{1}{BR}\sqrt{\frac{2Ve}{m}}$$

On squaring, $$\frac{e^2}{m^2} = \frac{1}{B^2R^2} \cdot \frac{2Ve}{m}$$

$$\frac{e}{m} = \frac{2V}{B^2 R^2} \quad ...(3)$$

By keeping B constant and varying V the ions of different atomic masses are made to describe the same circular orbit governed by eqn, (2).

VELOCITY SELECTOR

Let us consider that a charged particle is moving in crossed electric and magnetic fields of intensities $\vec{E}$ and $\vec{B}$ respectively. If $\vec{E}$ and $\vec{B}$ are acting along Y and Z axes respectively, if at any instant, the velocity of the particle, subjected to the two fields simultaneously, be $\vec{v} = \hat{i}v_x + \hat{j}v_y + \hat{k}v_z$. The velocity of charged particles along the z-axis the remains unaffected in the crossed magnetic and electric fields because neither the magnetic nor the electric field exerts any force on it. The magnetic field can only exert a force on a charged particle if it be moving in a direction perpendicular to the field, where a s the velocity of the particle, v_z, here in the direction of the field. And the electric field produces no effect on it because it is moving along the z-axis, *i.e.*, in a direction perpendicular to that of the field which is along the y-axis. The force acting on the particle due to the magnetic field will be because of its velocity v_x and will be equal to qv_xB along the negative direction of the y-axis *i.e.*, downward direction and the force acting on the particle due to the electric field E will be qE along the positive direction of the y-axis *i.e.*, upward direction. The magnetic field is so adjusted that the force qv_xB in the down direction due to it is balanced by the force in the upward direction and there will be no motion of the particle along the y-axis, *i.e.*, $v_y = 0$ if.

$$qv_x B = qE$$

Or $$v_x = E/B$$

i.e., if its velocity component along the x-axis be we equal to the ration between the electric and the magnetic fields. Since this expression for v_x is independent of mass and charge, this means that it applies equally well to tall charged particle irrespective of their masses and the sign and magnitude of the charges they carry.

This system of crossed field is known as a "velocity selector" of a "velocity filter" because it is used for selecting a beam of electrons or ions having constant velocity v = E/B and filtering out suitably adjusting the value of either field.

MAGNETRON

Valves in which the electrons move in a magnetic as well as in the electric field between the anode and cathode are called magnetron. Usually the electric and magnetic fields are perpendicular, such arrangement is called a plane magnetron. Cylindrical structures are generally employed. A cylindrical magnetron consists of a cylindrical abode along the axis of which is located the cathode. A radial electric field is set up by applying a high d.c. potential between cathode and anode. An axial magnetic field is provided by either permanent magnet or an electromagnet.

When magnetic field is not applied, the electrons will travel only due to electric field from cathode to anode.

With weak magnetic field a deflecting force acts at right angles to the B and to the direction of motion of the electrons and produce a circular motion. That is that electrons travel in a circular are between anode and cathode, the radius of curvature is given by

$$mv^2/R = eBv$$

or
$$R = mv/eB$$

The radius of curvature R decreases with increasing value of B. For a given value of v that corresponds to a given anode potential there is a 'critical value of magnetic field strength' which causes the electrons to just graze the anode .

This value of magnetic field strength is known as the "cut off value". If B exceeds the cut off value, the curvature of the electron path further decreases in such a way that electron misses the anode and spiral back to the cathode. Suppose the radius of anode be r_a and of cathode be r_k. The electrons are emitted from the surface of the cathode. An electron will experience an electrostatic force due to electric field intensity E as well as the magnetic force due to magnetic field B.

The force due to E is F = – eE and the force due to B is f = Bev where v is the velocity of electron.

But,
$$E = -dV/dr$$

hence
$$F = e\frac{dV}{dr}$$

and
$$f = Bev$$

The velocity v can be resolved into two components v_r along the radius and v_e perpendicular to the radius vector r. Therefore, the components of f are given by,

$$\left.\begin{aligned} f_r &= Bev_\theta \\ f_e &= Bev_r \end{aligned}\right\} \quad ...(1)$$

because magnetic force is ⊥ to the velocity v by relation $\vec{F} = e(\vec{v}\times\vec{B})$.

Let ω be the angular velocity and radius vector r makes an angle θ with an arbitrary line in the azimuthal plane, so that,

$$\omega = \frac{d\theta}{dt} = \frac{v_\theta}{r} \text{ since } v_\theta = r\frac{d\theta}{dt},, \ (v = \omega r)$$

Equation of motion is obtained by Newton's second law.

$$\frac{d}{dt}\left(m\frac{dr}{dt}\right) = F - f_r \quad ...(2)$$

The equation of motion in the azimuthal plane is obtained by equating the moment of impressed force to the rate of change of angular momentum.

$$rf_\theta = \frac{d}{dt}(mv_\theta . r) \quad ...(3)$$

Substituting f_θ and v_θ from above,

$$\frac{rBedr}{dt} = \frac{d}{dt}(mr^2\omega)$$

Integrating with respect to t, we have,

$$\int Ber\frac{dr}{dt}dt = \int\frac{d}{dt}\left(mr^2\omega\right)dt$$

$$\int Berdr = \int d\left(mr^2\omega\right)$$

$$Be\frac{r^2}{2} = mr^2\omega + C \quad ...(4)$$

The value of C is determined from initial condition, we know that when electron leaves the cathode *i.e.*, when

$$r = r_k, \ \omega = 0$$

$$C = \frac{Ber_k{}^2}{2} \quad ...(5)$$

therefore, $$\frac{Be}{2}\left(r^2 - r_k^2\right) = mr^2\omega$$

$$\omega = \frac{Be}{2}\left(1 - \frac{r_k^2}{r^2}\right) \quad ...(6)$$

Now if we ignore the small velocity with which the electron is emitted from the cathode the velocity v of the electron corresponding to the potential V is given by,

$$\frac{1}{2}mv^2 = eV$$

$$v = \sqrt{\frac{2eV}{m}}$$

When the current is just cut off, the electron is grazing the anode, then,

$$V = V_a$$

and $v_r = 0$ since there is no radial motion.

$$v_\theta = v = \sqrt{\frac{2eV_a}{m}}$$

$$r = r_a$$

therefore, $$\omega = v_\theta / r_a = \frac{1}{r_a}\sqrt{\frac{2eV_a}{m}} \quad ...(7)$$

If B_θ is the critical value of magnetic field corresponding to cut off value of anode current, then equating eqns. (6) and (7) replacing B in (6) by B_θ and r by r_a,

$$B_c \frac{e}{2m}\left(1 - \frac{r_k^2}{r_a^2}\right) = \frac{1}{r_a}\left(\frac{2eVa}{m}\right)^{1/2}$$

$$B_c = \left(\frac{8V_a}{e/m}\right)^{1/2} \frac{1}{r_a\left(1 - \frac{r_k^2}{r_a^2}\right)} \quad ...(8)$$

B_c is not depending on potential distribution. So that the presence of any other electrode or the space charge between the cathode and anode could not affect the cut-off conditions. The magnetron are of two types:

1. Split anode magnetron
2. Cavity magnetron

β-RAY SPECTROMETER

There are two chief methods for determining the energies of β-particles.

1. The magnetic spectrograph and
2. The Wilson cloud chamber.

The Magnetic Spectrograph Method

The basic principle of this method is based on the fact that when a charge particle enters the magnetic field then a force $\vec{F} = q(\vec{v}\times\vec{B})$ acts on it in a direction ⊥ to $\vec{v}$ and $\vec{B}$ hence the particle describes a circular path. The experimental technique in the case of β-rays are of three types depending upon the technique adopted for collecting the β-particles after they travel through the magnetic field.

1. The photographic plate technique
2. The Geiger tube counter technique
3. The Faraday cylinder technique

We shall discuss the 1st technique.

The photographic Plate Technique

In this process the β-rays given off by radioactive substances under study are bent by a constant magnetic field in such a way that they traverse semicircular path and then fall on photographic plate. Their motion is governed by,

$$\frac{mv^2}{R} = Bev \qquad \text{...(1)}$$

$$v = \frac{e}{m}BR. \qquad \text{...(2)}$$

β-particles of different velocities are there by focussed at different places on the plate. If the β-particles having velocities v_1, v_2,... they will move in circles of radii, R_1, R_2, ... governed by general relation (2). By suitably adjusting the position of the plate and the width of the slit S we can obtain 180° magnetic focussing. The β-particles of same velocity but slightly different initial directions being made to strike at the same point. The magnetic intensity used B can be measured. Using these values of R and B and assuming the value of e/m, the velocity v is calculated from relation (2). Hence, the kinetic energy $= \frac{1}{2}m\left(\frac{e}{m}BR\right)^2$ is known.

THE BETATRON

Principle

We have discussed that a time changing magnetic field produces an electrical field. When a loop of wire is placed in the changing magnetic field, an emf is induced in the loop that is equal to the line integral of the electric field around the loop (or circuit). Thus according to Faraday's law the induced emf is given by,

$$\xi_{in} = \oint \vec{E}.d\vec{l} = -\frac{d\phi_B}{dt}$$

by eqn. (11.1),

$$\xi_{in} = \oint \vec{E}.d\vec{l} = -\int_s \frac{\delta \vec{B}}{\delta t}.d\vec{S} \qquad ...(1)$$

When the loop is open this emf appears at the loop terminals.

It is highly significant to note that the induced electric field $\vec{E}$ *exists whether the wire loop is present of not. This fact is used for accelerating an electron and the device is called betatron.*

The betatron was developed by Donald W. Kerst about 1940, however, a device of this type having been proposed earlier by Max Steenback.

In a betatron electrons are accelerated by an electric field induced by time-changing magnetic field. The magnetic field is produced by an a.c. electromagnet with iron cone. Electrons are injected between the pole pieces of the magnet and are accelerated during a fraction of one a.c. period.

Let there be a magnetic field of flux density $\vec{B}$ directed out of the page from a pole piece. The time changing $\vec{B}$ induces an electric field $\vec{E}$. The force $\vec{F}$ on an electron due to this $\vec{E}$ is given be,

$$\vec{F} = e\vec{E} \qquad ...(2)$$

Since the electron is in a magnetic field, it tends to move in a circle, Thus, if $\vec{B}$ is outward as shown and increasing with time, the electron will rotate counter clockwise. The energy per unit charge acquired by an electron in travelling once around a circular orbit of fixed radius R is equal to the total induced emf around the path (by definition).

As $\vec{E}$ is everywhere tangential to the orbit therefore angle between $\vec{E}$ and $d\vec{l}$ s zero. Thus,

$$\xi_{ind} = \oint \vec{E}.d\vec{l} = E\oint dl$$

$$= E2\pi R = -\frac{d\phi_B}{dt}$$

where ϕ_B is total magnetic flux through the electron orbit (inside radius R). Disregarding the sign, the magnitude of the electric field is,

$$E = \frac{1}{2\pi R}\frac{d\phi_B}{dt}$$

substituting E in eq. (2) we get the magnitude of the force

$$F = eE = \frac{e}{2\pi R}\frac{d\phi_B}{dt}$$

By Newton's second law

$$F = \frac{d}{dt}(mv)$$

Equating these two expressions for force, we get,

$$\frac{d(mv)}{dt} = \frac{e}{2\pi R}\frac{d\phi_B}{dt} \qquad ...(3)$$

Suppose that the total flux ϕ_B through the electron orbit varies harmonically as given by,

$$\phi_B = \phi_{Bo} \sin \omega t$$

where ϕ_{Bo} = maximum magnetic flux

$$\omega = 2\pi f = \frac{2\pi}{T} = \text{radian frequency}$$

Electrons from a cathode may be injected into the betatron at a radius R at a time t = 0 when the flux ϕ_B is zero.

Then, if an electron is removed at some later time t less that one quarter period (T/4), its momentum is obtained by integrating eqn. (2). Therefore on integrating eqn. (3), we get momentum.

$$mv = \frac{e\phi_B}{2\pi R} \qquad ...(4)$$

where ϕ_B = total magnetic flux at time t through the electric orbit.

If we neglect the initial momentum of the electron, eqn. (4) gives the momentum mv of the electron at some time t after injecting the electron at t = 0. The velocity of the electron is

$$v = \frac{e\phi_B}{2\pi Rm} \text{ metre / sec.} \qquad ...(5)$$

However the radius R of the curved path in which electron moves is given by

$$mv^2/R = evB$$

or $$R = mv/eB \qquad ...(6)$$

where B = magnetic field at radius R.

Substituting mv from eqn. (4), in eqn. (6), we get,

$$B = \frac{e\phi_B}{2\pi eR^2} = \frac{\phi_B}{2\pi R^2}$$

But πR^2 = area (A) of electron orbit, therefore,

$$B = \frac{1}{2} = \frac{\phi_B}{A}$$

$$= \frac{1}{2} B_{av} \qquad ...(7)$$

where

B_{av} = average magnetic field inside electron orbit.

Both B and B_{av} are instantaneous values. Thus according to Eqn. (7), the instantaneous magnetic field B at the electron orbit (Radius R) must be one-half the average magnetic field inside the orbit a t the same instant when this condition is fulfilled, the radius R remains constant as the electrons are accelerated. *Therefore, the correct operation of a betatron requires that the average magnetic field inside the orbit increases at twice the rate of the magnetic field at the orbit itself.* In these circumstances, as the energy of the particle is increased by the induced electric field, the magnetic field at the orbit increases at just the rate required to keep the particle moving in a circle.

Construction

The betatron is so constructed that the condition given by eqn. (7) is satisfied. To obtain this the poles of the magnet can be given a proper shaping as indicated in the cross-sectional . This cross-section also shows the hollow, ring-shaped glass encloser, or doughnut, inside which the electrons travel.

To utilise the accelerated electrons, they may be made to impinge on a target and produce X-rays. The target is placed near the orbit radius

R, and a radial displacement of the orbit is caused by upsetting the 2 to 1 relation of B_{av} to B.

Advantage over Cyclotron

Betatron is free from the difficulty of relativistic variation of mass which prevents the use of the cyclotron for accelerating the electrons. As the betratron operates, the momentum of electron increases in proportionate to magnetic flux ϕ_B (or B_{av}) and if electron is to continue to move in its proper circle the relation (6) must continue to hold and B increases in proportional to B_{av} (or momentum mv) by equation (7).

Energy

can be estimated by knowing the average induced emf and total number or revolutions. The flux variation can be given by,

$$\phi_B = \phi_{Bo} \sin \omega t$$

The energy gained by the electrons per turn

$$e\, \xi_{ind} = e \frac{d\phi_B}{dt} = e\omega\phi_{Bo} \cos \omega t$$

The acceleration period is quarter of cycle *i.e.*,

T/4 where T = $2\pi/\omega$, hence acceleration period is

$$= \pi/2\omega \qquad ...(8)$$

The average value of the energy gained by electron per turn

$$= \frac{\text{Total energy during acceleration period}}{\text{Acceleration time}}$$

$$= \frac{e\omega\phi_{Bo} \int_0^{\pi/2\omega} \cos \omega t}{\pi/2\omega}$$

$$= \frac{e\omega\phi_{Bo} \left| \frac{1}{\omega} \sin \omega t \right|_0^{\pi/2\omega}}{\pi/2\omega}$$

$$= \frac{e\omega\phi_{Bo} \frac{1}{\omega} \sin \frac{\pi}{2}}{\pi/2\omega}$$

$$= \frac{2}{\pi} e\omega\phi_{Bo}$$

For most of the time electron travel with a velocity close to the velocity of light.

Therefore the distance travelled during the acceleration process

$$= c\,\frac{\pi}{2\omega}$$

The number of revolutions

$$= \frac{\text{Total distance travelled}}{\text{Distance travelled in one revolution}}$$

If R is radius of the orbit then in one revolution the electron travels a distance $2\pi R$.

Therefore the number of revolutions $= \dfrac{c\pi / 2\omega}{2\pi R}$

$$= \frac{c}{4\omega R}$$

Therefore the total energy gained by electron]

= Number of revolutions × mcan energy per turn

$$= e\omega\phi_{Bo} \cdot 2/\pi. = \frac{c}{4\omega R} = \frac{e\,c\,\phi_{Bo}}{2\omega R} \qquad ..(9)$$

However, the energy can also be estimated by the relativistics equation for energy,

$$\text{Kinetic energy} = pc$$

$$= BeRc \qquad ...(10)$$

SOLVED EXAMPLES

Example 1:

An electron has a velocity of 10^4 meters/sec. normal to a magnetic field of 0.1 weber/metre^2. Find the radius of the electron path and also it frequency.

Solution:

From equation (1), the radius,

$$R = \frac{9.1\times10^{-31}\times10^{4}}{1.6\times10^{-19}\times0.1} = 5.7\times10^{-7}\text{ metre}$$

This is a very small circle. The frequency.

$$n = \frac{v}{2\pi R} = \frac{10^4}{2\pi \times 5.7 \times 10^7}$$
$$= 2.8 \times 10^9 \text{ rPs.}$$

Example 2:

Find the frequency of the oscillator that is to be connected between the dees of a cyclotron with a magnetic field $\vec{B}$ = *1.5 weber/metre² for accelerating deutron. How the result is modified if we choose a-particle or proton?*

Solution:

$$q = 1.6 \times 10^{-19} \text{ couls.}$$

Mass of deutron = 2 × mass of proton

$$m = 3.34 \times 10^{-27} \text{ kg.}$$

From equation (1),

$$n = \frac{qB}{2\pi m}$$
$$= \frac{1.6 \times 10^{-19} \times 1.5}{2\pi \times 3.34 \times 10^{-27}} = 1.14 \times 10^7 \text{ RPS}$$
$$= 1.4 \text{ MC/sec.} \qquad (1 \text{ MC} = 10^6 \text{ cycles})$$

Accordingly, the oscillator frequency must be 11.4 MC/sec. If instead of deutron proton is accelerated then,

n = 22.8 MC/sec. since q/m for proton = twice of deutron

and for a = particle n = 11.4 M/C sec. as q/m for a particle = q/m of deutron.

Example 3:

In the above example, if the radius of the dees is 50 cm. find the energy of the emergent deutrons. Also calculate the energy of a = particle or proton if they were accelerated under the same condition.

Solution:

From equation (1),

$$W = \frac{1}{2}\frac{(RqB)^2}{m}$$

$$= \frac{(0.5\times 1.6\times 10^{-19}\times 1.5)^2}{2\times 3.34\times 10^{-27}}$$

$$= 2.15 \times 10^{-12} \text{ joule}$$

$$= \frac{2.15\times 10^{-12}}{1.6\times 10^{-19}}\text{eV}$$

$$= 1.33 \times 10^{7} \text{ eV}$$

$$= 1.33 \text{ MeV} \qquad (1 \text{ million electron volt} = 10^{6} \text{ eV})$$

If either a = particles or protons have been used the energy would be double, *i.e.*, 26.6 MeV because for a = particle q in the above example will be 2q and m will be 2m.

Hence,

$$W = \frac{4}{2}\left(\frac{1}{2}\frac{(RqB)^2}{m}\right) - 26.6\text{MeV}$$

Similarly for proton q s same. But mass is half of the deutron, *i.e.*, q = q, m = m/2,

$$W = 2\left[\frac{1}{2}\frac{(RqB)^2}{m}\right] = 26.6\text{MeV}$$

Example 4:

A stream of protons and deutrons in a vacuum chamber enters a uniform magnetic field Both are subjected to the same accelerating potential, hence kinetic energies are the same. If ion stream is perpendicular to the magnetic field and protons move in a circular path of radius 15 cm, find the radius of the path traversed by the deutrons. Given that mass of a deutron is twice that of a proton.

Solution:

The radius is given,

$$R = \frac{mv}{qB}$$

So that, radius of the path of proton

$$R_p = \frac{m_p v_p}{qB}$$

and radius of the path or deutron

$$R_d = \frac{m_d v_d}{qB}$$

Charge on deutron is same as on proton and moving in the same field. Thus,

$$\frac{R_d}{R_p} = \frac{m_d v_d}{m_p v_p}$$

Since they have same kinetic energies, *i.e.*,

$$\frac{1}{2} m_d \, v_d^{\,2} = \frac{1}{2} m_p \, v_p^{\,2}$$

or

$$\frac{v_d}{v_p} = \sqrt{\frac{m_p}{m_d}}$$

Since $m_d = 2\,m_p$, so

$$\frac{v_d}{v_p} = \sqrt{\frac{m_p}{2m_p}} = \frac{1}{\sqrt{2}}$$

Hence,

$$\frac{R_d}{R_p} = \frac{2m_p}{m_p} \cdot \frac{1}{\sqrt{2}} = \sqrt{2}$$

or $R_d = \sqrt{2}R_p$ since $R_p = 15$ cm (given)

hence $R_d = 0.15 \times \sqrt{2} = 0.212$ metre

$= 21.2$ cm.

Example 5:

A cyclotron accelerating protons has dees of 40 cm. radius and a radio frequency supply of 12 megacycles per second at 10000 volts maximum value. Calculate.

(1) The magnetic field B

(2) The speed and kinetic energy acquired by protons

(3) Maximum number of revolutions and time spent by a proton within dees.

Solution:

(1) From equation (1),

Cyclotron frequency $n = \dfrac{qB}{2\pi m}$

or $$B = \frac{2\pi mn}{q}$$

$$= \frac{2 \times 3.14 \times 1.67 \times 10^{-27} \times 12 \times 10^{6}}{1.6 \times 10^{-19}}$$

$$= 0.79 \text{ weber/metre}^2$$

(2) From equation (2),

Velocity acquired by proton,

$$v = \frac{qBR}{m}$$

$$= \frac{1.6 \times 10^{-19} \times 0.79 \times 0.40}{1.67 \times 10^{-27}}$$

$$= 3 \times 10^{7} \text{ metre/sec.}$$

Energy $= 1/2\ mv^2 = 1/2 \times 1.6 \times 10^{-27} \times (3 \times 10^7)^2$ joules

$= 7.5 \times 10^{-13}$ joules

$$= \frac{7.5 \times 10^{-13}}{1.6 \times 10^{-19}} \text{eV}$$

$$= 4.7 \times 10^{6} \text{ eV.}$$

Number of revolutions inside dees

$$N = \frac{WeV}{2qV} \times 1.6 \times 10^{-19}$$

$$N = \frac{4.7 \times 10^{6} \times 1.6 \times 10^{-19}}{2 \times \times 1.6 \times 10^{-19} \times 10{,}000}$$

$$= 235$$

Time spent within dees = 235/n

$$= \frac{235}{12 \times 10^{6}}$$

$$= 195 \times 10^{-5} \text{ sec.}$$

Example 6:

A cathode-ray tube with electrostatics deflection has an accelerating voltage, V_a = 1500 volts, length of deflecting plate l = 1 cm. Spacing between the plates (d) = 1 cm and screen is placed at a distance

L = 29.5 cm from the end of the deflecting plates. Find the voltage required to deflect the spot by 1 cm on the screen. Assuming that there is no fringing of the field.

Solution:

From eq. (1)

$$\frac{V_d}{Y} = \frac{2d\ V_a}{l(L+l/2)} = \frac{2\times10^{-2}\times1500}{10^{-2}\times(29.5+0.5)10^{-2}}$$

or
$$\frac{V_d}{Y} = 10{,}000 \text{ volts/metre}$$

or
$$\frac{V_d}{Y} = 100 \text{ volts/cm}$$

The required voltage is 100 volts.

Example 7:

A cathode-ray tube with magnetic deflection has an accelerating voltage V = 1500 volts, a magnetic deflecting field is uniform over length l = 2 cm and screen is placed at a distance L = 29 cm from the end of the deflecting field.

Find the magnetic field B required to deflect the spot of an electron beam 1 cm on the screen.

Solution:

From
$$\frac{B}{Y} = \frac{1}{l(L+l/2)}\sqrt{\frac{2m\ V_a}{e}}$$

For an electron on substituting e and m, we get,

$$\frac{B}{Y} = \frac{1}{l(L+l/2)}\ 3.38\times10^{6}\sqrt{V_a}$$

or
$$\frac{B}{Y} = \frac{3.38\times10^{6}\times1500^{1/2}}{2\times10^{-2}(29+1)10^{-2}}$$

or
$$\frac{B}{Y} = 2.18\times10^{-2} \text{ weber/metre}^2\text{/metre}$$

or
$$\frac{B}{Y} = 2.18\times10^{-4} \text{ weber/metre}^2\text{/cm}$$

Therefore the required magnetic field is 2.18×10^{-4} weber/metre^2.

Example 8:

Calculate the value of the electric field intensity which will give a proton an acceleration equal to the acceleration due to gravity.

Solution:

$$\text{Force} = eE = mg$$

or
$$E = mg/e$$

$$E = \frac{1.67\times10^{-27}\times9.8}{1.6\times10^{-19}} \text{ volt / cm}$$

$$= 10.2 \times 10^{-19} \text{ volt/cm.}$$

Example 9:

Two parallel plates are kept in air 1 cm apart and potential difference of 100 volt is applied. If there be an electron resting on the plate at lower potential. Calculate (1) the force on the electron, (2) the kinetic energy of this electron, when it reaches the other plate.

Solution:

$$E = \frac{V}{d}\ \frac{100}{0.01} = 10^4 \text{ volt / cm.}$$

Force
$$F = eE = 1.6 \times 10^{-19} \times 10^4$$
$$= 1.6 \times 10^{-15} \text{ Newton}$$

Acceleration

$$a = \frac{F}{m} = \frac{1.6\times10^{-15}}{9.1\times10^{-31}} = 1.75\times10^{15} \text{ metres / sec}^2.$$

velocity when it reaches the other plate,

v = Ö2ad Since $v^2 = u^2 + 2aS$ and

$$u = 0,\ S = d$$

or
$$v = \sqrt{2} \times 1.75 \times 10^{15} \times 0.01$$
$$= 5.9\times10^6 \text{ m/sec.}$$

Therefore, the kinetic energy

$$= \frac{1}{2}mv^2 = \frac{1}{2}\times9.1\times10^{-31}\times\left(5.9\times10^6\right)^2$$
$$= 1.6 \times 10^{17} \text{ joules.}$$

Example 10:

What must be the initial velocity of electrons which can just strike a conductor at – 1200 volts potential.

Solution:

The electrons, having negative charge, are repelled by a conductor at negative potential. In order that an electron just reaches the conductor, it should have the kinetic energy given by,

$$\frac{1}{2} m v^2 = eV$$

$$v = \sqrt{\frac{2eV}{m}} = \sqrt{\frac{2\times1.6\times10^{-19}\times1200}{9.1\times10^{-31}}}$$

$$= 2.06 \times 10^6 \text{ metre/sec.}$$

Example 11:

A positive ion of charge q and mass m is accelerated by a potential difference V, passes with a constant velocity through the space between two plates separated by distance d. Show that transit time between plates is $\frac{d}{(2qV/m)^{1/2}}$.

Solution:

From equation (1),

$$v = \sqrt{\frac{2qV}{m}}$$

If t be the transit line between the plates,

$$t = \frac{d}{v} = \frac{d}{\left(\frac{2qv}{m}\right)^{1/2}}$$

Example 12:

The electric field between the plates of a cathode-ray oscilloscope is 1.2 × 10⁴ Newton/coul. What deflection will an electron experience if it enters at right angles to the field with a kinetic energy of 2000 eV? The plates are 1.5 cm long.

Solution:

Using equation (1),

$$y = \frac{qE_y}{2m\,v_x^{\,2}}x^2$$

As given $1/2\ m\ v_x^{\,2} = 2000$ eV

$$y = \frac{qE_y x^2}{4\times\left(2000\times 1.6\times 10^{-19}\right)}$$

$$= \frac{1.6\times 10^{-19}\,\text{coul}\times 1.2\times 10^{4}\ \text{Nt/coul}\times(1.5\times 10^{-2}\,\text{metre})^2}{4\times 3.2\times 10^{-16}\,\text{joul}}$$

$$= 3.4 \times 10^4 \text{ metre}$$

$$= 0.34 \text{ mm}$$

Example 13:

An electron beam is accelerated through a potential difference of 1000 V. An electric field is applied at right angles to the path of the beam by means of rectangular parallel plates. The distance between the plates is 2.5 cm and potential difference 50 V. The edges of the plates are at a distance of 25 cm and 30 cm from the fluorescent screen. What is the displacement of the fluorescent spot ?

Solution:

$$E = \frac{50}{2.5\times 10^{-2}}\ \frac{\text{Volts}}{\text{metre}}$$

Kinetic energy $= 1/2\ mv^2 = eV$

$$1/2 \times 9.1 \times 10^{-31} \times v^2 = 1000 \times 1.6 \times 10^{-19}$$

On solving this for v, we get,

$$v = 5.6 \times 10^7 \text{ metre/sec}$$

Using equation (1).

$$y = \frac{qE}{m}\left(\frac{l}{2} + L\right)\frac{l}{v^2}$$

We have omitted the subscript and assumed that E is along y-axis and particle is moving along x-axis.

Substituting $L = 25 \times 10^{-2}$ metre

$$l = 5 \times 10^{-2} \text{ metre}$$

and v and E calculated above, we get,

$$y = \frac{50\times1.6\times10^{-19}}{2.5\times10^{-2}\times9.1\times10^{-31}}\left[0.25+\frac{0.05}{2}\right]\frac{5\times10^{-2}}{\left[5.65\times10^{7}\right]^{2}}$$

$$= 0.153 \times 10^{-2} \text{ metre}$$

$$= 0.153 \text{ cm.}$$

Example 14:

If a stream of electrons, each of mass m and charge e and velocity 3×10^6 metre/sec. is deflected by 2 mm in traversing a distance of 10cm through an electrostatics field of 18 V/cm. perpendicular to their path, find e/m in couls/kg.

Solution:

$$E = 18\,\frac{\text{Volts}}{\text{cm}} = \frac{18}{10^{-2}}\,\frac{\text{Volts}}{\text{metre}}$$

$$eE = ma$$

$$y = \frac{1}{2}at^2 = \frac{1}{2}\left(\frac{eE}{m}\right)\left(\frac{l}{v}\right)^2$$

$$2\times10^{-3} = \frac{1}{2}\left(\frac{18}{10^{-2}}\cdot\frac{e}{m}\right)\left(\frac{10\times10^{-2}}{3\times10^{6}}\right)^2$$

or
$$\frac{e}{m} = \frac{2\times10^{-3}\times2\times10^{-2}\times9\times10^{12}}{18\times100\times10^{-4}}$$

$$= 2 \times 10^{11} \text{ couls. kg.}$$

Example 15:

For a proton initially travelling with velocity vx = 5×10^8 cm/sec. Calculate the following:

(1) Force experienced in a magnetic field $\vec{B} = 2000\,\hat{j} + 4000\,\hat{k}$ *gauss*

(2) Acceleration experienced in an electric field E,

$$\vec{E} = 200\hat{i} + 100\hat{j} \text{ volts/cm}$$

(3) Transverse deflection in travelling a length L_x = 10 cm in an electric field E_y = 200 volts/cm.

Solution:

(1) Force in magnetic field

$$\vec{F} = q\left(\vec{v}\times\vec{B}\right)$$

$$\vec{v}_x = 5 \times 10^8 \text{ cm/sec} = 5 \times 10^6 \text{ metre/sec}$$

$$\vec{B} = 2000\ \hat{j} + 4000\ \hat{k}$$

$$\vec{B} = 0.2\ \hat{j} + 0.4\ \hat{k} \text{ weber/m}^2$$

Since, $\dfrac{1\text{ Weber}}{\text{metre}^2} = 10^4$ gauss

hence,

$$\vec{F} = 1.6 \times 10^{-19}\ [(5 \times 10^6\ \hat{i}) \times (0.2\ \hat{j} + 0.4\ \hat{k})]$$

$$= 1.6 \times 10^{-19} \times 5 \times 10^6 \times 2\ [0.1\hat{k} - 0.2\ \hat{j}]$$

$$= 1.6 \times 10^{-13}\ [\ \hat{k} - 2\hat{j}]\text{ Newtons}$$

Magnitude of the force F = 1.6×10^{-13} Ö5 = 3.57×10^{-13} Nt.

Its direction cosines

$$\left((0, -\frac{2}{\sqrt{5}}, \frac{1}{\sqrt{5}}\right)$$

(2) In electric field,

$$ma = qE$$

$$a = \frac{qE}{m} = \frac{1.6\times10^{-19}}{1.67\times10^{-27}}\left(\frac{200\hat{i}+100\hat{j}}{10^{-2}}\right)$$

$$= \frac{1.6}{1.67}10^{12}\ (2\hat{i}+\hat{j})$$

$$= \frac{1.6}{1.67}10^{12}\sqrt{5}$$

$$= 2.14 \times 1012 \text{ met/sec}^2$$

(3) Transverse deflection,

$$Y = \frac{1}{2}at^2$$

$$= \frac{1}{2}\left(\frac{gE_y}{m}\right)\left(\frac{1}{v_x}\right)^2,$$

$$\text{Since } E_y = 200\ \frac{\text{volt}}{\text{cm}} = \frac{200}{10^{-2}}\ \frac{\text{volt}}{\text{metre}}$$

$$Y = \frac{1}{2} \times \frac{1.6 \times 10^{-19} \times 200}{1.67 \times 10^{-27} \times 10^{-2}} \times \left(\frac{10 \times 10^{-2}}{5 \times 10^{-6}}\right)^2 \text{ metre}$$

$$= 0.0383 \times 10^{-2} \text{ metre}$$

$$= 0.0383 \text{ cm}$$

Example 16:

An electron of energy 20 eV moving in a direction perpendicular to a magnetic field B = 10^2 weber/m^2 describes a circle or radius R. Calculate the value of R.

Solution:

Kinetic energy $\frac{1}{2} mv^2 = 20 \text{ eV}$

$$= 20 \times 1.6 \times 10^{-19} \text{ joules}$$

hence, $$v = \left(\frac{2 \times 20 \times 1.6 \times 10^{-19}}{m}\right)^{1/2} \text{ m/sec}$$

The radius of the circular path is given by equation (1),

$$R = \frac{mv}{qB}$$

$$= \frac{m \times \left(64 \times 10^{-19}\right)^{1/2}}{\sqrt{m \times q \times B}}$$

$$= \frac{\left(9 \times 10^{-31} \times 64 \times 10^{-19}\right)^{1/2}}{1.6 \times 10^{-19} \times 10^2}$$

$$= 15.0 \times 10^{-8} \text{ metre}$$

Example 17:

An electron of velocity $\vec{v} = (2\hat{i} + 3\hat{j})\,10^8$ cm/sec. enters a region of uniform magnetic field, $B = 500\,\hat{i}$ gauss, show that its path becomes helical.

(1) In what direction does the axis of the helix lie ?

(2) Calculate the radius of helix

(3) Calculate the number of rotations as the electron advances 10 cm along the axis of the helix.

Solution:

$\vec{v} = (2\hat{i} + 3\hat{j})\ 10^8$ cm/sec. and

$\vec{B} = 500\hat{i}$ gauss

The component of velocity $\vec{v}_1 = 2\hat{i} \times 10^8$ cm/sec. is along $\vec{B}$ and electron moves along x-axis and due to the component $\vec{v}_2 = 2\hat{j} \times 10^8$ cm/sec. which is perpendicular path.

(1) The axis of helix lines along B *i.e.*, x-axis.

(2) Radius of helix R by equation (1)

$$R = \frac{mv_2}{qB} + \frac{9\times10^{-31}\times3\times10^6}{1.6\times10^{-19}\times500\times10^{-4}}$$

(1 Weber/metre^2 = 10^4 gauss)

$$R = 3.37 \times 10{-}4 \text{ metre}$$

$$= 0.0337 \times 10{-}2 \text{ metre}$$

$$= 0.0337 \text{ cm}$$

(2) Distance travelled in one rotation = path of the helix by equation (1),

$$\frac{2\pi m\, v_1}{qB} = \frac{2\pi\times9\times10^{-31}\times2\times10^6}{1.6\times10^{-19}\times500\times10^{-4}}$$

Number of rotations made by the electron as it advances through a distance of 10 cm

$$= \frac{10\times10^{-2}}{\text{pitch of helix}}$$

$$= \frac{10\times10^{-2}\times1.6\times10^{-19}\times500\times10^{-4}}{2\pi\times9\times10^{-31}\times2\times10^6}$$

$$= 70.74 = 71.$$

EXERCISES

1. The uranium isotopes of atomic weights 235 and 238 may be separated by 180° magnetic focussing. Calculate the separation of their beams at the focus, brought in magnetic field of 10^4 gauss

if the radius of curvature of the heavier isotopes be 100cm when (a) the velocities, (b) the energies of the beams are equal.

2. Explain with a diagram the construction and action of a cathode ray oscillograph. Give an account of its practical applications.
3. Obtain expressions for deflection sensitivity of an electrostatic CRO and magnetic CRO. Compare the two results.
4. (a) What do you meant by time base circuit?

   (b) Describe the procedure for measuring phase difference by CRO.

   (c) Show how you can study the hysteresis loop of a specimen of ferro-magnetic material with CRO.
5. Describe J.J. Thomson's method for determining e.m of an electron.
6. Draw detailed diagram to illustrate the principle of Thomson's mass spectrograph. What are the limitations of this method ?
7. Show that a charged particle is moving in crossed electric and magnetic field perpendicular to both the fields can go undeflected if,

$$V = E/B$$

8. Describe with necessary theory Aston's mass spectrograph. Discuss how it has been used for the detection of isotopes.
9. Describe 180° magnetic focussing.
10. Show that gain in kinetic energy of a charged particle in an electric field is equal to qV where V is the potential difference between its initial and final positions. Define an electron volt.
11. Show that the path of charged particle moving with a uniform initial velocity in a constant transverse electric field is a parabola. Find out an expression for the direction of emergence of the particle from the field with initial direction of motion.
12. Explain clearly the principle underlying the working of a cyclotron. What are its drawbacks and how they have been modified?
13. Show that the cyclotron frequency of a given kind of ions is independent of the energy. State the limiting conditions, if any.
14. Describe Dampster's mass spectrograph and show how it is used to separate the isotopes.

15. Give a detailed account of Bainbridge's mass spectrograph.
16. What is Bainbridge's velocity selector device ?
17. Mention different types of mass spectrographs and their merits and demerits.
18. Calculate the intensity of electric field which would produce in an electron an acceleration equal to that due to gravity.
19. An electron beam is accelerated through a potential difference of 1000 volts and then passed through a pair of deflecting plates 2 cm long and 0.5 cm apart. What should be the potential difference between the plates in order that the electron beam may be deflected through 5°.
20. An electron describes a helix of radius 6 cm and pitch 2.0 cm in a magnetic field of intensity 30 gauss. Obtain the component of its velocity along and perpendicular to the magnetic field.

5

Magnetic Scalar Potential

INTRODUCTION

In the same year, Biot and Savart formulated the equation for the field due to current in a long straight wire. Finally, they proposed a mathematical formula for the field due to a single current element rather than a finite length of wire, the equation can be used to calculate the field due to many different current configurations as shown in Fig. 5.1. In fact, we are often interested not in the field of moving charge, but in that of an element of current, as a length dl of wire carrying a current I. We can easily set up an equivalence with equation (1). Suppose cross-section of wire is A and ρ is the charge density of charge in wire. Assume this charge moves with velocity v. The first significant relationship between a current and its magnetic field was discovered in 1820 by Oersted who found that a magnetised needle tends to set itself at right angle to a straight wire in which there is a steady current.

The charge crossing any cross-section per second = ρvA that is equal to current I

or $$\rho vA = I.$$

If we multiply by dl on both sides, we get,

$$\rho vAdl = Idl$$

$$qv = Idl \qquad ...(1)$$

(ρAdl is the total charge contained in wire of length dl.)

So the magnetic induction $d\vec{B}$ due to dl length of wire carrying current I is given by the following relation at a point P shown in Fig. 5.1. After substituting qv = Idl in equation (1), we get,

$$d\vec{B} = \frac{\mu_o}{4\pi} I \left(\frac{\vec{dl} \times \hat{r}}{r^2} \right) \quad ...(2)$$

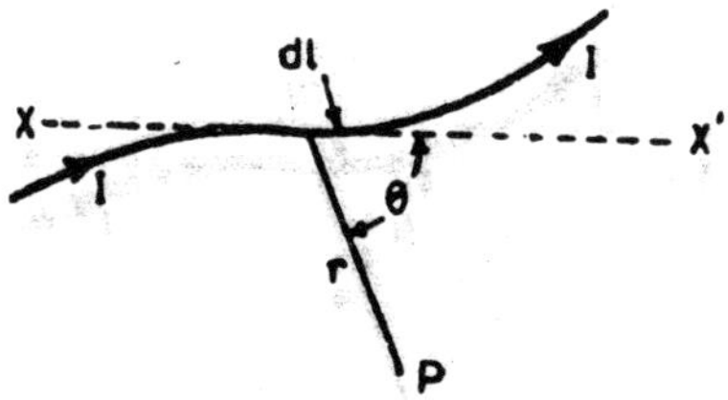

Fig. 5.1

or,

$$d\vec{B} = \frac{\mu_o}{4\pi} I \left(\frac{\vec{dl} \times \vec{r}}{r^3} \right) \quad ...(2a)$$

The law described by equations (1) and (2) is called Biot-Savart law.

Principle of Superposition

The resultant field at P is found by integrating equation (4.2) as the principle of superposition also holds here as in the case of electric field,

$$\vec{B} = \oint d\vec{B}$$

or

$$\vec{B} = \frac{\mu_o}{4\pi} \oint \frac{I}{r^3} \left(\vec{dl} \times \vec{r} \right) \quad ...(3)$$

where $\oint$ stands for close path.

The direction of $d\vec{B}$ is normal to the line from the current element Idl to the observation point P. (By the rule discussed in last chapter).

Right Hand Rule

Grasp the wire with right hand, the thumb pointing in the direction of current, the fingers will curl around the wire in the direction of $\vec{B}$.

THE FORCE BETWEEN TWO FINITE ELEMENTS OF CURRENT

Consider two current elements $I_1 dl_1$ and $I_2 dl_2$ flowing in different conductors as shown in Fig. 5.2. The magnetic induction due to I_2 current at a distance r on dl_1 is,

$$d\vec{B} = \frac{\mu_o}{4\pi} I_2 \left(\frac{\vec{dl}_2 \times \vec{r}}{r^3} \right)$$

Now the force on conductor carrying current I_1, say dF, due to current I_2 can be calculated by using the relation $(\vec{F} = I\vec{l} \times \vec{B})$.

Thus the force between two elements of current $I_1 dl_1$ and $I_2 dl_2$ is given by,

$$d\vec{F} = \frac{\mu_o I_1 I_2}{4\pi} \left(\frac{d\vec{l}_1 \times (d\vec{l}_2 \times \vec{r})}{r^3} \right) \qquad ...(1)$$

This law for current element is equivalent to Coulomb's law of force between static charges.

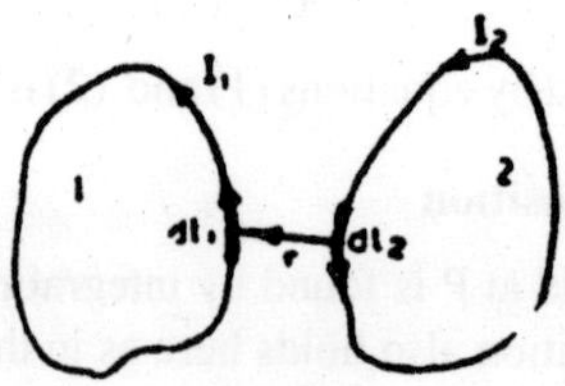

Fig. 5.2

Biot-Savart Law and Coulomb's Law of Electrostatics

Eqs. 1 and 2 reveal that Biot-Savart law is similar to Coulomb's law in electrostatics. The field or force varies inversely as the square of the distance from the current element. It differs from Coulomb's law, since the direction of field is normal to the line from the current element (Idl) to the point of observation, instead of being directed along this line, s in Coulomb's law.

APPLICATIONS OF BIOT-SAVART LAW

we shall calculate the B in some important cases as given below:

Magnetic Field due to Steady Current in a Long Straight Wire

As an example of use of Biot-Savart law, let us calculate the magnetic field vector $\vec{B}$ at a point P a distance R from a long straight wire in which there is a current I as shown in Fig. 5.3. The magnetic field $d\vec{B}$ due to current element $Id\vec{l}$ (*i.e.*, dl length of the wire carrying current I) of the wire at point P is given by Biot-Savart law *i.e.*,

$$d\vec{B} = \frac{\mu_o}{4\pi} I \left(\frac{d\vec{l} \times \vec{r}}{r^2} \right) = \frac{\mu_o}{4\pi} \frac{Idl}{r^2} (\hat{j} \times \hat{r})$$

where $\hat{j}$ is unit vector along y-axis and

$\hat{r}$ is unit vector along r.

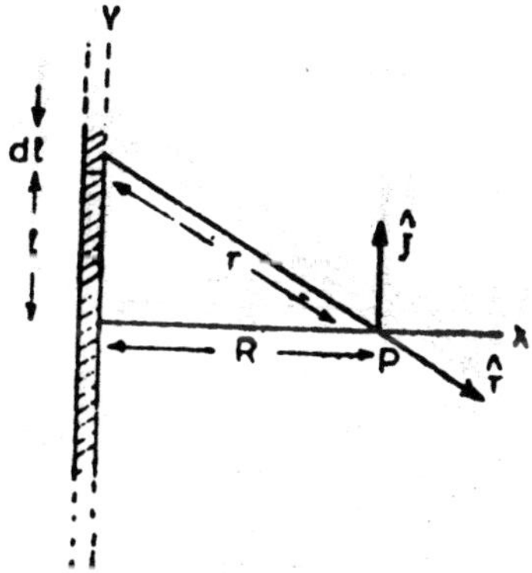

Fig. 5.3

or $$dB = \frac{\mu_o}{4\pi} \frac{Idl \sin\theta}{r^2}$$

The direction of $d\vec{B}$ at point P for all elements are the same, that is, into the plane of the figure at right angle to the page. The magnitude of $d\vec{B}$ due to whole wire is given by,

$$B = \int dB = \frac{\mu_o I}{4\pi} \int_{l=-\infty}^{l=+\infty} \frac{\sin\theta \, dl}{r^2}$$

From Fig 5.3,

$$r = (l^2 + R^2)^{1/2}$$

and , $$\sin\theta = \sin(\pi - \theta) = R/r$$

$$= \frac{R}{(l^2 + R^2)^{\frac{1}{2}}}$$

Putting these values, the expression for B becomes,

$$B = \frac{\mu_o I}{4\pi} \int_{-\infty}^{+\infty} \frac{R dl}{(l^2 + R^2)^{\frac{3}{2}}} \quad ...(1)$$

To evaluate the integral on R.H.S., substitute,

$$l = R \tan \phi, \qquad dl = R \sec^2 \phi \, d\phi$$

therefore,
$$B = \frac{\mu_o I}{4\pi R} \int_{-\pi/2}^{+\pi/2} \cos \phi \, d\phi$$

$$= \frac{\mu_o I}{2\pi R} \qquad ...(2)$$

The direction of $\vec{B}$ is specified by $(\hat{j} \times \hat{r}) = \hat{\theta}_o$ and is perpendicular to the plane containing $\hat{j}$ and $\hat{r}$. Hence,

$$\vec{B} = \frac{\mu_o I}{2\pi R} \hat{\theta}_o \qquad ...(2a)$$

Also the direction of $\vec{B}$ by R.H. rule i.e, grasp the wire with right hand, the thumb pointing in the direction of current, the finger will curl around the wire in the direction of B, hence magnetic field lines are circular closed curves around the wire.

Magnetic Field Near a Straight Current Filament of Finite Length

Let us calculate the magnetic field due to a wire of length $2l$ at a point P as shown in Fig. 5.4. The geometry and notations to be used are shown in Fig. 5.4. The wire is placed along x-axis and origin of the coordinate system is at the middle. Further, let the point P be at a distance R from the x-axis.

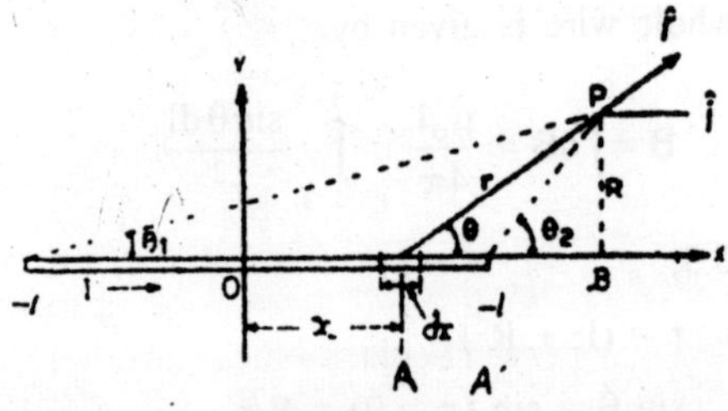

Fig. 5.4

Consider the contribution to the magnetic field by the current element Idx at P as shown in Fig. 5.4. According to equation (2(a)) the magnetic field $d\vec{B}$ is given by,

$$d\vec{B} = \frac{\mu_o}{4\pi} I \left(\frac{d\vec{x} \times \hat{r}}{r^2} \right)$$

Form figure $r^2 = AB^2 + R^2$,

and $AB = AA' + A'B$

where $AA' = l - x$ and

$A'B = R \cot \theta_2$

Hence $AB = l - x + R \cot \theta_2$

So $r2 = (l - x + R \cot \theta_2)^2 + R^2$, therefore,

$$d\vec{B} = \frac{\mu_0}{4\pi} \frac{Idx}{(l - x + R \cot \theta_2)^2 + R^2} (\hat{i} \times \hat{r})$$

where, $d\vec{x} = \hat{i}dx$, and

$\hat{i}$ = unit vector along x-axis

From the figure, it is seen that in triangle APB,

$$l - x + R \cot \theta_2 = R \cot \theta$$

then $dx = R \operatorname{cosec}^2 \theta \, d\theta$

and $\operatorname{cosec} \theta = \frac{r}{R} \frac{\sqrt{(l - x + R \cot \theta_2)^2 + R^2}}{R}$

and $dx = \frac{(l - x + R \cot \theta_2)^2 + R^2}{R} d\theta$

Substituting dx in the expression $d\vec{B}$, we get,

$$d\vec{B} = \frac{\mu_0}{4\pi} \frac{I}{R} d\theta \left(\hat{i} \times \hat{r}\right)$$

But the magnitude of the product $\hat{i} \times \hat{r}$ is $\sin \theta$ and direction is perpendicular to plane containing $\hat{i}$ and $\hat{r}$,

i.e., $\hat{i} \times \hat{r} = \hat{\theta}_o \sin \theta d$

where $\hat{\theta}_o$ is unit vector perpendicular to the plane containing $\hat{i}$ and $\hat{r}$, hence,

$$d\vec{B} = \frac{\mu_0}{4\pi} \frac{I}{R} \sin \theta \, d\theta \, \hat{\theta}_o \qquad ...(2b)$$

Magnitude $dB = \frac{\mu_0 I}{4\pi R} \sin \theta \, d\theta$

The total field is obtained by integrating between the limit sql and θ_2 due to length 2l of the wire,

$$B = \int_{\theta_1}^{\theta_2} dB \quad \int_{\theta_1}^{\theta_2} \frac{\mu_o I}{4\pi R} \sin\theta \, d\theta$$

$$= \frac{\mu_o I}{4\pi R} \left| -\cos\theta \right|_{\theta_1}^{\theta_2}$$

$$B = \frac{\mu_o I}{4\pi R}(\cos\theta_1 - \cos\theta_2)$$

and directed along $\hat{\theta}_o$, hence,

$$\vec{B} = \frac{\mu_o I}{4\pi R}(\cos\theta_1 - \cos\theta_2)\,\hat{\theta}_o \qquad \text{...(2c)}$$

Consequently, the magnetic field is circular and by symmetry has its centre on the axis. For the special case when the length becomes very long, there by approaching the infinite case, *i.e.*,

$$l \to \infty, \ \theta_1 \to 0, \text{ and } \theta_2 \to \pi$$

then,

$$\bar{B} = \frac{\mu_o I}{2\pi R}\hat{\theta}_o \qquad \text{...(2d)}$$

which is same as given in equation (2(d))

Interaction Between Two Parallel Long Currents

Long parallel wires separated by distance d and carrying currents I_a and I_b. It is an experimental fact, noted by Ampere only one week after Oersted's magnetic phenomena of electric current. Wire a will produce a field of induction B_a at all nearby points. The magnitude of B_a, due to current I_a at a distance d *i.e.*, on wire b is,

$$B_a = \frac{\mu_o I_a}{2\pi d}$$

According to the R.H. rule the direction of B_a is downward as shown in Fig. 5.5. Wire b, which carries a current I_b is now really placed in such field. Consider the l length of the wire b it will experience a force $(I\vec{l} \times \vec{B})$ whose magnitude is,

$$F_b = I_b l B_a = \frac{\mu_o l I_a I_b}{2\pi d}$$

(After substituting Ba given above) ...(3)

Direction of F_b is determined according to vector rule F_b lies in the plane of the wires and points to the left.

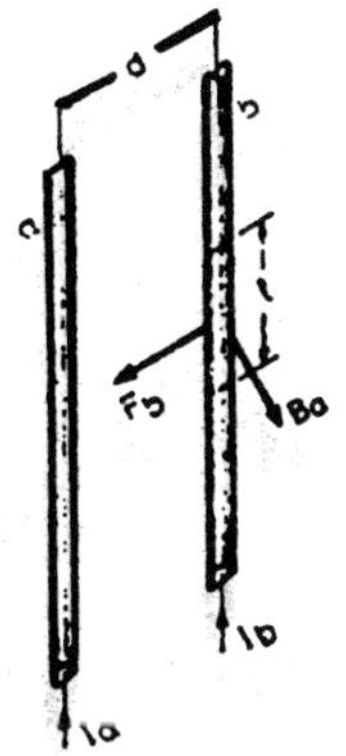

Fig. 5.5

One can start with wire b, compute the field of induction which it produces at the site of wore a, and then compute the force on wire a, *for parallel currents, F_a points to the right. The forces that the two wires exert on each other are equal and opposite, hence according to Newton's* law of action and reaction, *they will attract. For antiparallel currents, the two wires repel each other.* We can compare the above discussion with that of two point charges. There we explained that two charges act on each other through the intermediary of the electric field. Here two conductors in same manner by; magnetic field. We think as,

$$\text{current} \rightleftarrows \text{field}$$

and not as in the action at a distance point of view in terms of,

$$\text{current} \rightleftarrows \text{current.}$$

Unit of Current in Rationalised MKS

The force of attraction between two very long parallel wires is used to define the ampere practically. The ampere is that current which when flowing through two long parallel straight conductors placed in vacuum at a distance of one metre from each other produces a force of attraction of 2×4^{-7} Newtons per metre length between the wires, *i.e.*, when in equation (4),

$$l = 1 \text{ metre, } I_a = I_b = 1 \text{ amp, d} = 1 \text{ metre}$$

$$F = 2 \times 10^{-7} \text{ Newtons}$$

Hence $\mu_o = 4\pi \times 10^{-7}$ Newtons/amp^2.

$$= 4\pi \times 10^{-7} \frac{\text{Newtons. metre}}{\text{Ampere}^2\text{. metre}}$$

$$= 4\pi \times 10^{-7} \frac{\text{Weber}}{\text{Amp. metre}}$$

Since, $$1 \text{ Weber} = \frac{1 \text{ Newton. metre}}{1 \text{ Ampere}}$$

'Basically' the ampere is defined by equation (5). Thus if we take two current elements each of unit length, and at unit distance apart, so oriented that the factor,

$$\frac{d\vec{l}_1 \times \left(d\vec{l}_2 \times \vec{r}\right)}{r^3} = 1$$

Then the currents are defined to an ampere, *i.e.*,

$$I_1 = I_2 1 \text{ amp}$$

if the resulting force is 10^{-7} Newtons. From this it follows by equation (5),

$$\mu_o = 4\pi \times 10^{-7} \text{ Newton/ampere}^2$$

$$\mu_o = 4\pi \times 10^{-7} \text{ Weber/amp.metre}$$

Be Careful to Note The Following Facts

The value assigned to μ_o is seen to be the result of definition rather than experiment. This, on the other hand is not the case with ε_o, which we encountered in Coulomb's law of electrostatics. The difference is that we define th unit of current, the ampere, from the Biot-Savart law, whereas, we define the coulomb, the unit of charge, as the charge flowing per second in a current of 1 ampere. Therefore, we are not free to choose the constant *i.e.*, ε_o in Coulomb's law at our own choice but must determine it by experiment.

Magnetic Field Along The Axis of Circular Coil

Consider a circular coil of radius R and carrying a current I. The vector $d\vec{l}$ for a current element at the top of the coil points perpendicular towards the reader. The $d\vec{l}$ and $\vec{r}$ are perpendicular to each other and plane formed by $d\vec{l}$ and $\vec{r}$ is normal to the page as shown in Fig. 5.6.

The $d\vec{B}$ for this element is at right angle to this plain, this lies in the plane of the figure and perpendicular to r. The magnitude of dB from equation (6) is given by

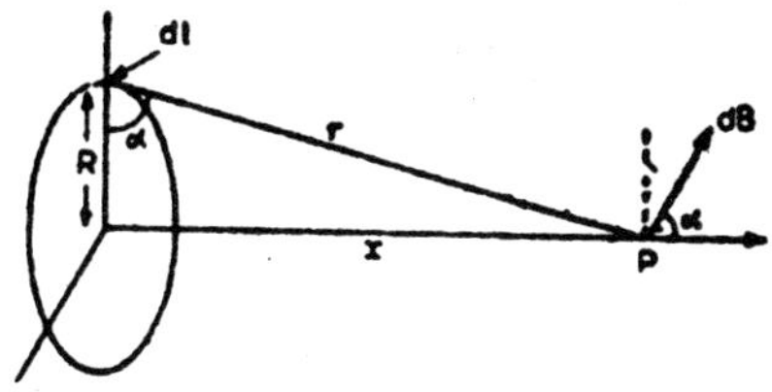

Fig. 5.6

$$dB = \frac{\mu_o}{4\pi} \frac{Idl}{r^2} \qquad ...(6)$$

Let us resolve $d\bar{B}$ into two components, one, $d\vec{B}\|$, along the axis of loop and another, $dB\perp$,at right angles to x-axis. As the coil is symmetrical about the x-axis the contribution $d\bar{B}$ due to element of coil on opposite side the $dB\perp$ will be equal in magnitude but opposite in direction and therefore, cancel out, we have only $dB \|$ component. The resultant B for the complete loop is given by,

$$B = \int dB = \int \frac{\mu_o I}{4\pi R} dl \cos\alpha$$

Since $dB\| = dB \cos\alpha$ and dB is given in Eq. (7) and from figure.

$$r = \sqrt{R^2 + x^2}$$

$$\cos\alpha = R/r = R/\sqrt{R^2 + x^2}$$

therefore,

$$B = \int \frac{\mu_o I}{4\pi} \frac{Rdl}{(R^2 + x^2)^{3/2}} \qquad ...(7)$$

$$= \frac{\mu_o IR}{4\pi(R^2 + x^2)^{3/2}} \int dl$$

$\int dl$ is the circumference of the loop $2\pi R$.

$$B = \frac{\mu_o IR^2}{2(R^2 + x^2)^{3/2}} \qquad ...(8)$$

If coil having N turns, then

$$B = \frac{\mu_o NIR^2}{2(R^2 + x^2)^{3/2}} \qquad ...(8a)$$

This shows the B decreases as the distance x increases. At the centre $x = 0$

$$B_{centre} = \frac{\mu_o IR^2}{2R^3} = \frac{\mu_o I}{2R} \quad ...(9)$$

If x >> R *i.e.*, point P lies far from coil in Eq. (8), R in denominator can be neglected. Hence,

$$B = \frac{\mu_o IR^2}{2x^3} \quad ...(10)$$

If coil having N turns, then,

$$B = \frac{\mu_o NIR^2}{2x^3} \quad ...(11)$$

Recalling the area of the coil is πR^2,

$$B = \frac{\mu_o (NIA)}{2\pi x^3}, \ (A = \pi R^2)$$

and NIA according to definition given in the last magnetic dipole moment of the current coil represented by m

$$B = \frac{\mu_o m}{2\pi x^3} \quad ...(12)$$

Variation of The Along The Axis of Coil and Principle of Helmholtz Galvanometer

Differentiating Eqn. (8) with respect to x, we get,

$$\frac{dB}{dx} = \frac{\mu_o IR^2}{2} (-3/2).\ 2x(R^2 + x^2)^{-5/2}$$

$$\frac{d^2B}{dx^2} = -\frac{3\mu_o IR^2}{2} [(R_2 + x^2)^{-5/2} + x(-5/2).\ 2x\ (R^2 + x^2)^{-7/2}]$$

or, $$\frac{d^2B}{dx^2} = -\frac{3\mu_o IR^2}{2} (R^2 + x^2)^{-7/2} [R^2 + x^2 - 5x^2]$$

$$\frac{d^2B}{dx^2} \text{ is zero when } R^2 + x^2 - 5x^2 = 0$$

or $$x = \pm \frac{R}{2} \quad ...(13)$$

and at those points $\frac{d\vec{B}}{dx}$ is constant. Hence, at a distance of $\frac{R}{2}$ on either side of the coil, the rate of change of the field becomes constant. This is the principle of Helmholtz galvanometer. The variation is shown in Fig. 5.7.

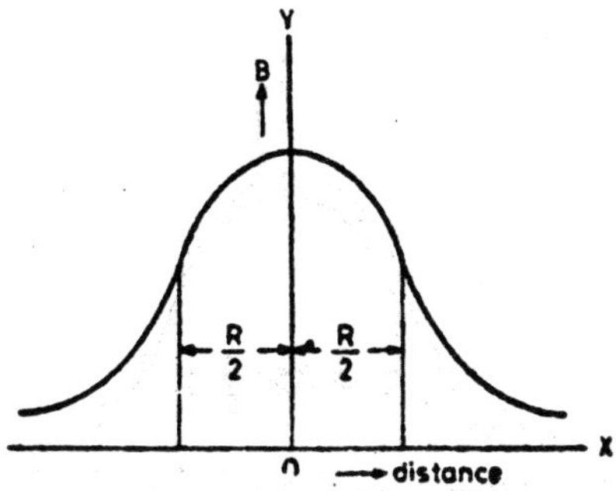

Fig. 5.7

Field on Axis of Solenoid

The result of Eqn. (14) may be extended to find vector $\vec{B}$ at a point P on the axis of a solenoid. A coil of wire wound in the form of cylinder is called solenoid, Fig. 5.8(a). Assuming the uniform winding, so that the number of terms per unit length of the solenoid are constant, say, n. Let us first calculate the field due to a small segment of length dx of the solenoid which subtends angle θ and θ + dθ with the axis at the point P. From Fig. 5.8(b) we have,

$$\frac{rd\theta}{dx} = \sin\theta \quad \text{or} \quad dx = \frac{rd\theta}{\sin\theta}$$

The number of turns on the length dx $= \dfrac{nrd\theta}{\sin\theta}$

and the total current in this element dx $= \dfrac{Inrd\theta}{\sin\theta}$

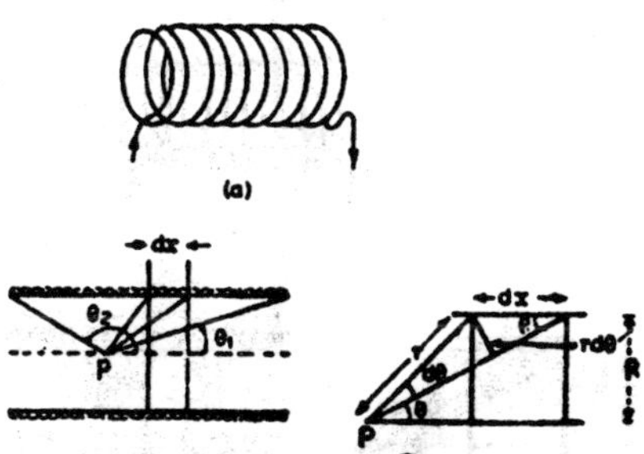

(a) (b)

Fig. 5.8

where I is the current flowing in the solenoid putting this n Eqn. (14) in place of I we get the contribution of the current rings in a length dx of the solenoid. The direction of the field is along the axis of coil that is x-axis hence dB due to dx element is given by,

$$dB = \frac{\mu_0 R^2}{2r^3} \frac{Inrd\theta}{\sin\theta}, \text{ since } \sin\theta = R/r$$

we get

$$= \frac{\mu_0 In \sin\theta \, d\theta}{2}$$

The field due to whole solenoid is

$$B = \frac{\mu_0 In}{2} \int_{\theta_1}^{\theta_2} \sin\theta \, d\theta$$

$$= \frac{\mu_0 In}{2} (\cos\theta_1 - \cos\theta_2) \quad ...(15)$$

Note that if the fingers of the right hand are curled around the solenoid in the direction of circulations of the current, the thumb will point in the direction of B.

If the solenoid is very long and P is some where near the middle, $\theta_1 = 0$, and $\theta_2 = \pi$ then,

$$B = \mu_o In \quad ...(16)$$

If the point P lies at one end of the solenoid, $\theta_1 = 0$, and $\theta_2 = \pi/2$ and thus,

$$B = \mu_o n I/2 \quad ...(17)$$

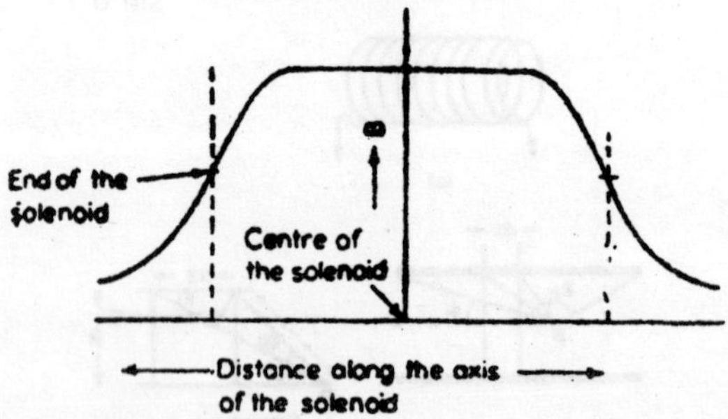

(c)

Fig. 5.8

This shows that $\vec{B}$ *at the end of a long solenoid is just one half of that at the centre,* the variation is shown in Fig. Fig. 5.8 and it is clear that B at the centre of the solenoid remains nearly constant until we reach one of the ends.

The Biot-Savart law can be written as in the case of closed current loop,

$$\vec{B} = \frac{\mu_o I}{4\pi} \oint \frac{d\vec{l} \times \hat{r}}{r^2}.$$

We now wish to show that the magnetic field near a current loop is proportional to the gradient of the solid angle subtended by the loop at the point where $\vec{B}$ is to be found. Let us displace the point P in Fig. 3.1 by an infinitesimal distance $d\vec{r}$ and find the value of $\vec{B}.\, d\vec{r}$

$$\vec{B}.\, d\vec{r} = \frac{\mu_o I}{4\pi}\left(\oint \frac{d\vec{l} \times \hat{r}}{r^2}\right).\, d\vec{r}$$

The value of R.H.S. remains the same whether we first evaluate the integral then take the dot product with $d\vec{r}$ or we first take the product of $d\vec{l} \times \hat{r}/r^2$ with $d\vec{r}$ and then evaluate the integral around the whole circuit. Hence we can write,

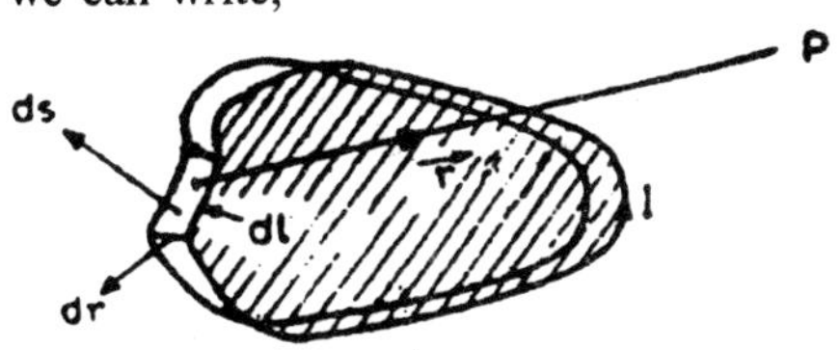

Fig. 5.9

$$\vec{B}.\, d\vec{r} = \frac{\mu_o I}{4\pi} \oint \frac{(d\vec{l} \times \hat{r}).d\vec{r}}{r^2}$$

By the rules of triple scalar product *i.e.*,

$$(\vec{A} \times \vec{B}).\, \vec{C} = (\vec{B} \times \vec{C}).\, \vec{A} = (\vec{C} \times \vec{A}).\vec{B}$$

$$(d\vec{l} \times \hat{r}).\, d\vec{r} = (d\vec{r} \times d\vec{l}).\, \hat{r}$$

Therefore, $$\vec{B}.\, d\vec{r} = \frac{\mu_o I}{4\pi} \oint \frac{(d\vec{r} \times d\vec{l}).\hat{r}}{r^2}$$

$$= \frac{\mu_o I}{4\pi} \oint \frac{(d\vec{l} \times (-d\vec{r})).\hat{r}}{r^2}$$

$$(\because \vec{A} \times \vec{B} = -\,\vec{B} \times \vec{A})$$

Now. $d\vec{l} \times (-d\vec{r}).\ \hat{r}/r^2$ *can be interpreted in terms of solid angles. The interpretation is clear, when the point P is displaced by an amount dr, the solid angle* Ω *subtended by the circuit at P changes by* $d\Omega$. *If we keep P fixed and move the entire current or in other words the each point of the circuit is displaced by an amount – dr, even then the change in solid angle is* $d\Omega$.

The quantity $d\vec{l} \times (-d\vec{r})$ is a vector representing the small area $d\vec{S}$ shown as a ribbon surface in Fig. 3.2 *i.e.*, $d\vec{S} = d\vec{l} \times (-d\vec{r})$. The projection of this area on a plane normal to r is given by

$$d\vec{l} \times (-d\vec{r}).\ \hat{r}$$

Therefore,
$$\frac{d\vec{l} \times (-d\vec{r}).\ \hat{r}}{r^2} = \frac{d\vec{S}.\hat{r}}{r^2} = d\Omega$$

where $d\Omega$ represents the element of solid angle subtended by the area dS at P. Thus, the integral over the closed loop gives the entire solid angle subtended by ribbon. This solid angle is shown in Fig. 3.2. Here, it represents the decrease of the entire solid angle Ω subtended at P, that occurs when P moves a distance $d\vec{r}$ away from the loop moves – $d\vec{r}$ away from P Therefore, we write,

$$\oint \frac{d\vec{l} \times (-d\vec{r}).\ \hat{r}}{r^2} = -\, d\Omega.$$

Since Ω is a scalar function,

$$d\Omega = \vec{\nabla}\ \Omega.d\vec{r}$$

therefore,
$$\oint \frac{d\vec{l} \times (-d\vec{r}).\ \hat{r}}{r^2} = -\ \nabla\ \Omega.d\vec{r}$$

Hence,
$$\vec{B}.\ d\vec{r} = -\frac{\mu_o I}{4\pi} \nabla\ \Omega.d\vec{r}$$

or
$$\vec{B} = -\frac{\mu_o I}{4\pi}\ \nabla\ \Omega = -\ \nabla\left(\frac{\mu_o I\Omega}{4\pi}\right) \qquad ...(18)$$

This Eqn. is the solid angle form of Biot-Savart law or the Ampere's law. The direction of $\vec{B}$ is that of – $\nabla\ \Omega$, so that $\vec{B}$ points away from the loop along its positive normal Eqn. (18) is the solid angle form of Ampere law. Now as we know that Ω is a scalar function. It is clear from eqn. (1) that the magnetic induction $\vec{B}$ is expressed in terms of scalar function Ω (x, y, z) of the coordinate of P which is analogous to the relation of electrostatics,

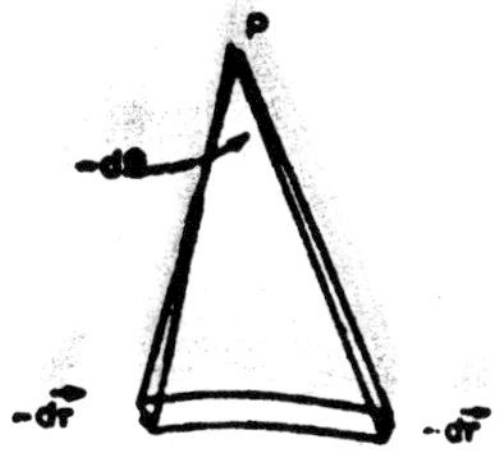

Fig. 5.10

$$\vec{E} = -\nabla V$$

We can write Eqn. (19) as

$$\vec{B} = -\nabla V_m \quad ...(19)$$

where,

$$V_m = \mu_o I \Omega / 4\pi$$

$$= \frac{\mu_o}{4\pi} \times \text{current} \times \text{solid angle} \quad ...(20)$$

and for collection of loops,

$$V_m = \frac{\mu_o}{4\pi} \sum_K I_k \; \Omega_k \quad ...(21)$$

V_m is a scalar potential whose negative gradient gives the magnetic induction $\vec{B}$. This is generally called a 'scalar potential' since it is scalar quantity, to distinguish it from the vector potential, which we shall soon introduced. It is clear that, in cases, in which $\vec{B}$ can be derived from a scalar potential, we must have curl $\vec{B} = 0$, since the curl of any gradient of scalar is zero.

Comment on Scalar Potential

The electrostatic potential is single valued. However, the potential V_m is no single-value since the solid angle is not single-valued. In face solid angle is not a single-valued function of position P. If point P makes a series of circuits through the loop, solid angle (Ω) will change to $\Omega + 4\pi n$ where n is some positive or negative integer. Unlike electrostatic potential, the difference between the scalar potential between any tow points is also undefined unless the path followed between the two points is specified. This is so because each time, the path links the circuit, the solid angle along the path changes by 4π and the difference in magnetic scalar potential changes by,

$$\frac{\mu_o}{4\pi}(\Omega + 4\pi - \Omega) = \mu_o I.$$

Thus, the magnetic scalar potential v_m is multiple valued. *This method may be used for calculating the potential of conductors of negligible cross-sections. It cannot be applied for points within the conductors themselves.* It can be used for points well outside the current circuit under consideration. We can choose the value of Ω and hence of V_m which goes to zero at infinity. Hence V_m can be defined by an equation similar to the electric potential with reference point chosen at infinity.

$$V_m = -\int_{\infty}^{P} \vec{B}.d\vec{l}$$

Unit

From relation $Vm = \mu_o I\ \Omega/4\pi$, the MKS unit has dimensions of μ_o times amperes. *The unit for V is not completely given.*

We have not interpreted V_m as representing energy for the reason that B does not directly represent a force on a real entity. If single magnetic pole really existed then $\vec{B}$ would be the force on unit pole at the point in question. Hence we say Vm as a sort of fictional potential energy per unit pole.

THE MAGNETIC DIPOLE

Suppose we have a current I flowing in a positive direction around a very small loop of area A, the normal to the loop being $\hat{n}$. If θ is the angle between $\hat{n}$ and the vector from the loop to a point P, as shown in Fig. 5.11, the solid angle subtended by A at P will be $A\cos\theta/r^2$, so that the scalar potential will be by eqn. (1)

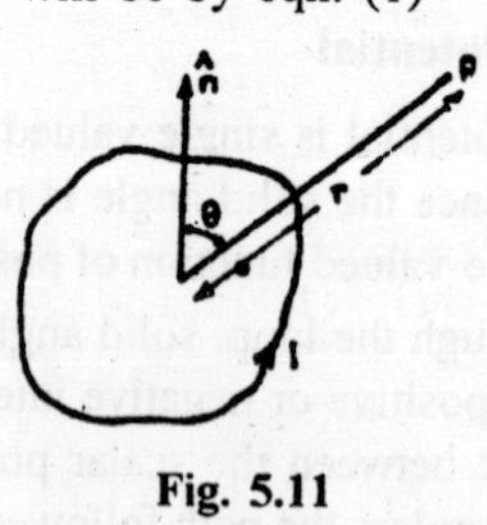

Fig. 5.11

$$\frac{\mu_o I}{4\pi}\ \frac{A\cos\theta}{r^2}$$

This expression depends on position of the point P, just like the potential of an electric dipole,

$$V = \frac{1}{4\pi\varepsilon_o}\ \frac{P\cos}{r^2} = \frac{1}{4\pi\varepsilon_o}\frac{\vec{p}.\hat{r}}{r^2}$$

Thus, a small loop carrying a current has magnetic induction $\vec{B}$, whose lines of force are like those of an electric dipole, we call such a small loop a magnetic dipole and define the product IA as the dipole moment. If this moment is represented by m, we have,

$$\text{Potential of magnetic dipole} = \frac{\mu_o m\cos\theta}{4\pi r^2} = \frac{\mu_o}{4\pi}\ \frac{\vec{m}.\hat{r}}{r^2}$$

$$= \frac{\mu_o}{4\pi}\ \frac{\vec{m}.\hat{r}}{r^3} \qquad ...(2)$$

Magnetic Shell and its Equivalence with a Current Loop

In terms of this formulation, there is an interesting way in which we can regard the magnetic field of a current loop as carrying from a distribution of magnetic dipole over a surface spanning the loop shown in Fig. 5.12, we may divide the surface into may small loops, let a current I flow in each of them in positive direction. These currents will cancel each other on all the interior boundaries of the loops, but not over the outer boundary. Thus, the sum of the currents of all the small loops will give just the current originally present in the large loop. So that the magnetic field of the small loops must equal that of large loop. Each of the small loops of area A, however, produce a field like a dipole of moment IA. In other words, a distribution of dipoles over surface, whose dipole moment per unit area is I, distributed over any surface spanning the current loop will have the same magnetic induction at all points that the current loop itself has.

Strength of Magnetic Shell

It may be defined as the magnetic moment per unit area. Therefore, strength of the magnetic shell = IA/ A = I

Potential at a Point Very Close to the Surface

$$V_m = \frac{\mu_o I}{4\pi} \times \text{solid angle}$$

in this case solid angle = 2π.

Hence, $V_m = \frac{\mu_o I}{4\pi} I \, 2\pi = \frac{\mu_o I}{2}$ ...(3)

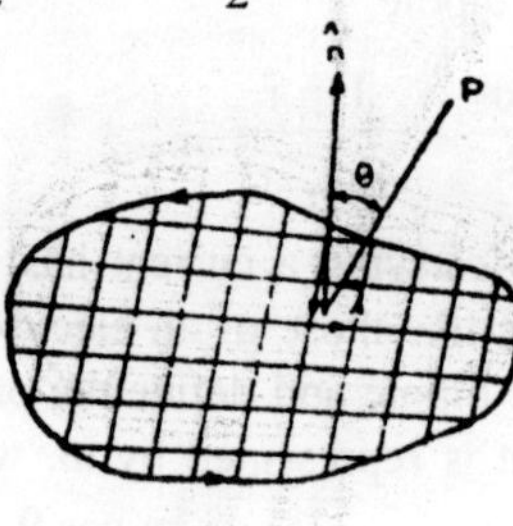

Fig. 5.12

At a Point Inside a Closed Magnetic Shell,

Solid angle = 4π

Hence, $V_m = \mu_o I$ ...(4)

AMPERE'S CIRCUITAL LAW OR AMPERE'S LAW

Ampere by using relation t = m B sing θ, in which t is torque, m is the magnitude of the magnetic moment and θ is the angle between the direction of B and normal to the wire, measured B by bringing a compass needle near a straight wire.

Even though m for a compass needle is not known, m may be regarded as constant independent of the position or orientation of the needle. Thus by measuring t and q one can obtain a relative measure of B for various distances r and for various currents I in the wire. The experimental results can be represented mathematically,

$$B \propto I/r.$$

We can convert this proportionality into an equality by inserting a proportionality constant $\mu_o/2\pi$ with the same reasoning as far Coulomb's law μ_o is permeability constant for free space

$$B = \frac{\mu_o I}{2\pi r}$$

or $B\,(2\pi r) = \mu_o I$

The L.H.S. is $\oint \vec{B}.d\vec{l}$ for a path consisting of a circle of radius r centred on the wire. For all points on this circle $\vec{B}$ has the same constant magnitude of $\vec{B}$ and $d\vec{l}$ which is always tangent to the path of integration,

$$\oint \vec{B}.d\vec{l} = \oint B.dl = B\oint dl = B(2\pi r)$$

In this special case we can write the experimentally observed connection between $\vec{B}$ and I, $\oint \vec{B}.d\vec{l} = \mu_o I$ ...(1)

Statement of Ampere's Law

The law states that the line integral of $\vec{B}$ around a close path is equal to μ_o times the current enclosed by the path if the path does not enclose the current then,

$$\oint \vec{B}.d\vec{l} = 0.$$

PROOF OF AMPERE'S LAW–THE LINE INTEGRAL OF THE MAGNETIC FIELD VECTOR

We can prove the Ampere's law which is a fundamental relation between the current and the line integral of the magnetic field vector.

Line Integral for Irregular Path Which Encloses Current

Now we shall examine the line integral in the close path of integration around the current wire of an irregular shape as shown in Fig. 5.13(a). First determine the value of line integral along a small section dl of total path, which is shown in amplified form in Fig. 5.13(b). $\vec{B}$ along the are $d\vec{l}$ is given in the similar way as above.

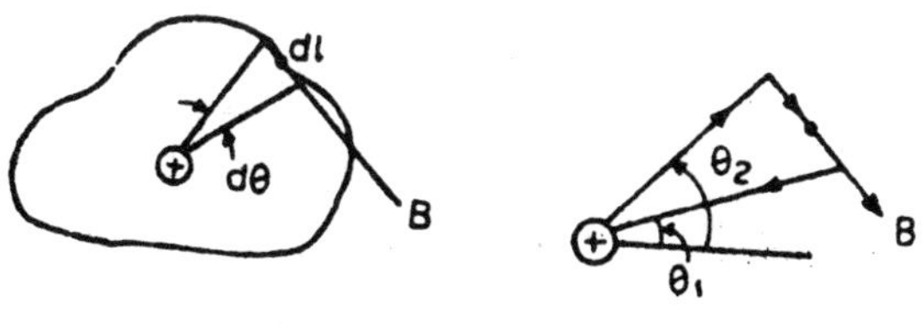

(a) (b)

Fig. 5.13

$$\int_{are} \vec{B}.d\vec{l} = \frac{\mu_o I}{2\pi}\int_{\theta_1}^{\theta_2} d\theta = \frac{\mu_o I}{2\pi}(\theta_2 - \theta_1)$$

Along the radial path $\vec{B}$ and $d\vec{l}$ are perpendicular to each other where $\vec{B}.\ d\vec{l}$ is zero. So

$$\int_{\substack{radial\\path}} \vec{B}.d\vec{l} = 0.$$

Clearly for the element of the path shown, therefore,

$$\int_{path} \vec{B}.d\vec{l} = \frac{\mu_o I}{2\pi}(\theta_2 - \theta_1)$$

Since the result is independent of the distance R, then by taking the summation of the contribution to the total integral by all segments into which the total path has been divided, it is clear that $\theta_2 - \theta_1 = 2\pi$ for closed path that encloses the current

$$\oint \vec{B}.d\vec{l} = \mu_o I.$$

If the path is taken either radial or parallel to the current, $\vec{B}$ and $d\vec{l}$ are normal to each other. Since any arbitrary $d\vec{l}$ can be analysed into radial, tangential and parallel components, it is clear that only the tangential component contributes to the integral. Therefore, one chooses an arbitrary path for the integration around the current. The path could again be divided into segments as above only tangential components will be considered and the same result.

Value of $\oint \vec{B}.d\vec{l}$ When Path does not Enclose Current

Here too the path which does not enclose the current is broken into segments as shown in Fig. 5.14(a). Based on ;the same consideration as in the case when path enclosed the current for the path shown in Fig. 5.14(b) two tangential contributions exist.

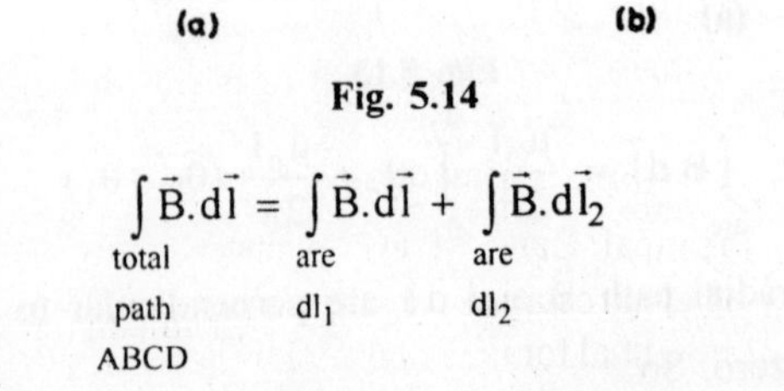

Fig. 5.14

$$\text{Thus,} \quad \int_{\substack{total\\path\\ABCD}} \vec{B}.d\vec{l} = \int_{\substack{are\\dl_1}} \vec{B}.d\vec{l} + \int_{\substack{are\\dl_2}} \vec{B}.d\vec{l}_2$$

But the two integrals on R.H.S. have equal and opposite values. However the value of integral vanishes for BC and AD, B and dl are perpendicular, hence $\vec{B}.d\vec{l} = 0$. For complete path, we have

$$\int_{ABCD} \vec{B}.d\vec{l} = \int_A^B \vec{B}.d\vec{l} + \int_B^C \vec{B}d\vec{l} + \int_C^D \vec{B}.d\vec{l} + \int_D^A \vec{B}.d\vec{l}$$

$$= \int_A^B \vec{B}.d\vec{l} + \int_C^D \vec{B}.d\vec{l}$$

$$= \int_A^B \vec{B}.d\vec{l}_2 - \int_D^C \vec{B}.d\vec{l}_1 \text{ (since } dl_2 \text{ and } dl_1 \text{ are opposite)}$$

B at a distance $r_1 = \dfrac{\mu_0 I}{2\pi r_1}$

B at a distance $r_2 = \dfrac{\mu_0 I}{2\pi r_2}$

Hence,
$$\int_{ABCD} \vec{B}.d\vec{l} = \frac{\mu_0 I}{2\pi r_2}\int_A^B dl_2 - \frac{\mu_0 I}{2\pi r_1}\int_D^C dl_1$$

$$= \frac{\mu_0 I}{2\pi r_2} AB - \frac{\mu_0 I}{2\pi r_1} DC$$

Now, from Fig. 4.20 (b),

$$DC/AB = r_1/r_2$$

or
$$DC = (r_1/r_2)\, AB$$

Putting this value of DC in the above equation, we get,

$$\int_{ABCD} \vec{B}.d\vec{l} = \frac{\mu_0 IAB}{2\pi r_2} - \frac{\mu_0 I r_1 AB}{2\pi r_1 r_2} = 0$$

Thus, the results can be summarised,

$$\oint \vec{B}.dl = \mu_0 I \quad \text{if path links current} \qquad ...(1)$$

$$= 0 \quad \text{if path does not link current}$$

Ampere's Law and Gauss's Law of Electrostatics

Eqn. (1) is identical Gauss's law. Ampere's law is identical to Gauss's theorem of electrostatics for calculating magnetic field. As in electrostatics the differential form of Gauss's law is a fundamental equation of Maxwell electromagnetic theory. Similarly, Ampere's law will be also expressed as a fundamental equation of electromagnetic theory we will just see it. However, we can apply the Ampere's law where $\oint \vec{B}.d\vec{l}$ is easily evaluated *i.e.*, in problems having sufficient symmetry.

APPLICATIONS OF AMPERE'S LAW?

When do We Apply Ampere's Law

Like Gauss's law, Ampere's law is *always true* (for steady currents), but it is not always useful. The usefulness of Ampere's law depends on out ability to evaluate $\oint \vec{B}.d\vec{l}$ For this, we construct a closed path, called Amperian path near the current-distribution, for which the magnitude of field and angle between $\vec{B}$ and to the direction of path, have constant value; since then B-cos θ can be taken outside the integral sign. B is constant for uniform and symmetrical current-distribution and θ is constant for symmetrical path. *Thus, symmetry is crucial for the application of Ampere's law.* Therefore, we apply Ampere's law only when we have uniform and symmetrical, current-distribution and symmetrical Amperian path. The standard current-configurations which can be handled by Ampere's law are :

1. infinite straight lines
2. infinite planes
3. infinite solenoids
4. toroids.

When Ampere's law applies, it's by far the fasted method; when it doesn't, we'll have to fall back on the Biot-Savart law.

APPLICATIONS OF AMPERE'S LAW

We shall apply this law for calculating B in important cases given below:

1. $\vec{B}$ Near a Long Wire

As we know that the line of magnetic induction for a long straight wire carrying a current I are concentric circles centred on the wire. The central dot in Fig. 5.15 represents a current I in the wire emerging from the page. The angle between $\vec{B}$ and $d\vec{l}$ is zero, hence $\vec{B}.\,d\vec{l}$ = BDl, then applying the Ampere's law We have, $\vec{B}$ at point P at a distance r

$$\oint \vec{B}.d\vec{l} = \mu_o I$$

$$B.2\pi r = \mu_o I$$

$$B = \frac{\mu_o I}{2\pi r} \quad ...(1)$$

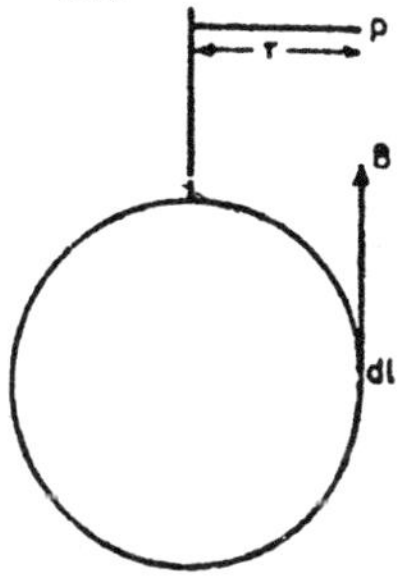

Fig. 5.15

In fact we may regard this as an experimental result consistent and also derivable from Ampere's law. This expression is same as obtained by Biot-Savart law.

It is interesting to compare Eqn. (1) with the expression for electric field near a long line of charge,

$$E = \frac{I}{2\pi\varepsilon_o} \frac{\lambda}{r}$$

Both are identical depending on constant multiple factor $\mu_0/2\pi$ and $1/2\,\pi\varepsilon_0$ and the factor responsible for the fields I and λ. Finally each field varies as $1/r$. They are derived in the same manner $\vec{E}$ by using Gauss's law and evaluating integral over Gaussian surface and $\vec{B}$ by using Ampere's law and evaluating for closed circular path. However, lines of $\vec{E}$ are perpendicular to the surface, while lines of $\vec{B}$ are everywhere tangent to the circular path.

2. Two Parallel Conductors

Whether we use Biot-Savart law or Ampere's law, B near a long wire at a distance r from the wire given by $\mu_0 I/2\pi r$. Hence, we can explain the force between the two currents in the same manner as discussed previously.

3. B for a Solenoid

A solenoid is a long wire wound in a close-packed helix and carrying a current I. For points near to a single turn, the behaviour of the wire is same as that of straight wire. The total field is the vector sum of the fields set up by all the turns. B is parallel to the solenoid axis at the interior points.

The field at external point is zero. For points such as P the field set up by the upper part of the solenoid turns marked O points to the left and tends to cancel the field set-up by the lower part of the solenoid turns mark ⊗, which points to the right as shown in Fig. 5.16(a).

Let us apply Ampere's law,

$$\oint \vec{B}.d\vec{l} = \mu_0 I$$

to the path abcd shown in Fig. 5.16. The total integral is sum of four integrals, one for each path of rectangle,

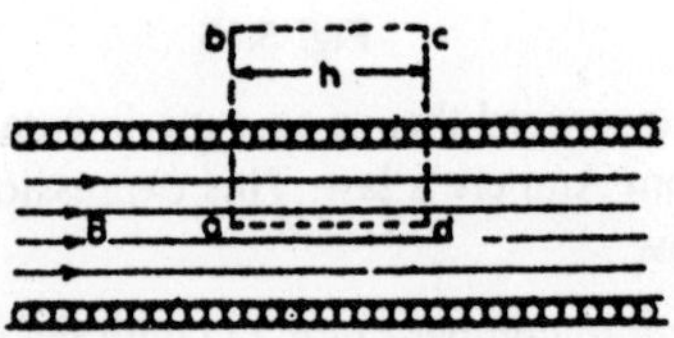

Fig. 5.16

$$\int_{abcd} \vec{B}.d\vec{l} = \int_a^d \vec{B}.d\vec{l} + \int_d^c \vec{B}.d\vec{l} + \int_c^b \vec{B}.d\vec{l} + \int_b^a \vec{B}.d\vec{l}$$

for ba, B and dl are perpendicular to each other hence $\vec{B}.\ d\vec{l} = 0$ and similarly for dc $\vec{B}.\ d\vec{l} = 0$. The path cb is outside the solenoid $\vec{B}.\ d\vec{l}$ is zero because B is zero outside the solenoid. The only contribution to the field is due to the path ad thus for entire path,

$$\int_{abcd} \vec{B}.d\vec{l} = \int_a^d \vec{B}.d\vec{l}$$

B and dl are parallel inside, hence,

$$\int_{abcd} \vec{B}.d\vec{l} = B\int_a^d dl, \text{ since } \int_a^d dl = \text{lenght ad} = h$$

$$= Bh$$

The net current I that passes through the area bounded by the path of integration is not the same as the current I_o in the solenoid because path of integration encloses more than one turn.

Let n be the number of turns per unit length, then

$$I = J_o(nh)$$

So, $$Bh = \mu_o I_o nh \qquad ...(2)$$

The relation is obtained for an infinitely long ideal solenoid, it holds for actual solenoid for internal points near the centre of the solenoid. The relation does not depend on the diameter or the length of the solenoid and B is constant over the solenoid cross section. If N are the total number of turns and l is the length of wire of the solenoid, then

$$n = N/l$$

$$B = \mu_o I_o (N/l) \qquad ...(3)$$

same expression as derived by Biot-Savart law.

4. A Toroid

Fig. 5.17 shows a toroid. A toroid may be visualised as a doughtnut with wire would uniformly around the doughnut. The toroid may be considered as a solenoid that has been bent into a circle with the ends joined. Since the axis of the resulting bent solenoid is circular, the lines of $\vec{B}$ are also circular.

Let us apply Ampere's law to the circular path of integration of radius r,

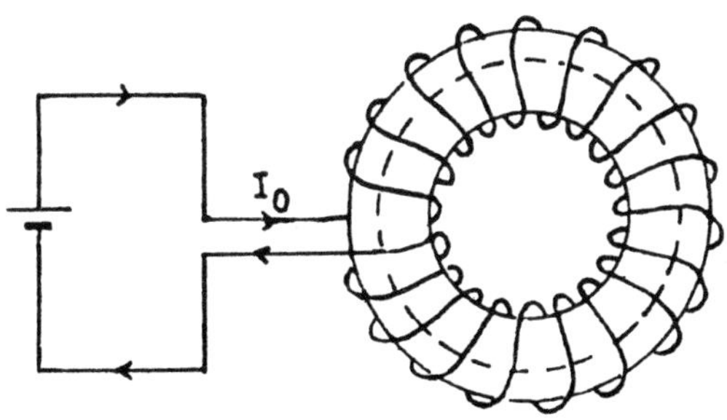

Fig. 5.17

$$\oint \vec{B}.d\vec{l} = \mu_o I$$

$$B.2\pi r = \mu_o I_o N$$

where I_o is the current in the toroid windings and N are the total number of turns. Therefore,

$$B = \frac{\mu_o}{2\pi} \frac{I_o N}{r} \qquad ...(4)$$

In contrast to the solenoid, B is not constant over the cross-section of a toroid.

For External Point

As discussed in the case of solenoid the field at outside point is zero. So here also B is zero because toroid is a bent circular solenoid.

CHARACTER OF $\vec{B}$ LINES AND THE DIVERGENCE OF $\vec{B}$

In our study of electrostatics, we first found simple methods of obtaining the field by integrating the contribution resulting from all points charges according to Coulomb' law. We soon found, however that this was method of restricted usefulness. The really powerful tools of electrostatics come only when we apply general analytical methods, introducing the potential, Gauss's theorem, Poisson's equation, and similar relations. Here too, we shall introduce general analytical methods. We shall first show the div. $\vec{B}$ is always zero.

The lines of magnetic field are closed circle around the axis of a current element. In other words, the lines of magnetic field produced by currents always close on themselves. This means of course, that if an infinitesimal region of space were examined, the total number of $\vec{B}$ lines entering the specified region must be equal to the total number of $\vec{B}$ lines leaving the specified region. As we have already established for steady current where the continuity prevails for the number of lines of electric currents is characterised mathematically by div.$\vec{J}$ = 0. Same condition is satisfied by vector $\vec{B}$ as $\vec{B}$ lines are continuous, hence we must have div. $\vec{B}$ = 0.

To be more specific, since the magnetic field lines do not have sources in contrast to the case of electric field where the lines emerge from positive charges and end on the negative charges, as we know the divergence of a vector signifies the starting or stopping of the vector, we should expect that div.

$$\vec{B} = 0 \qquad ...(1)$$

Proof:

We can prove div.$\vec{B}$ = 0 from Biot-Savart law and using vector analysis. Noting that for small element of current $I d\vec{l} = \vec{I} dl$. According to Biot-Savart law, the field of complete circuit is given by,

$$\vec{B} = \frac{\mu_o}{4\pi} \int \frac{\vec{I}\, dl \times \hat{r}}{r^2}$$

taking the divergence of both sides,

$$\text{div} \cdot \vec{B} = \nabla.\vec{B} = \nabla.\left(\frac{\mu_0}{4\pi}\int\frac{\vec{I}\, dl\times\hat{r}}{r^2}\right)$$

$$= \frac{\mu_0}{4\pi}\int\nabla\left(\frac{\vec{I}\times\hat{r}}{r^2}\right)dl$$

Since, it makes no difference whether we first evaluate integral then take divergence or take the divergence first then evaluate the integral.

therefore $$\text{div}.\ \vec{B} = \frac{\mu_0}{4\pi}\int\nabla.\left(\frac{\vec{I}\times\hat{r}}{r^3}\right)dl$$

Since $$\nabla\left(\frac{I}{r}\right) = -\ 1/r^2$$

hence $$\text{div}.\ \vec{B} = -\frac{\mu_0}{4\pi}\int\nabla.\left[I\times\nabla\left(\frac{I}{r}\right)\right]dl$$

But according to the vector analysis expansion,

$$\nabla.(\vec{A}\times\vec{B}) = \vec{B}.\nabla\times\vec{A} - \vec{A}.\nabla\times\vec{B}]$$

therefore, $$\text{div}.\ \vec{B} = -\frac{\mu_0}{4\pi}\int\left[\nabla\left(\frac{I}{r}\right).\nabla\times\vec{I} - \vec{I}.\nabla\times\nabla\left(\frac{I}{r}\right)\right]dl$$

To interpret this result it is recalled that $\vec{B}$ is being specified at the field point, where as the current is that at the source points specified by dl. Say the field point is specified by the variable. (x, y, z) the source point might be specified by the variable (x', y', z') But since the divergence operation is being taken with respect to the field point variables which are independent of the source point variables, then

$$\nabla\left(\frac{I}{r}\right).\nabla\times\vec{I} = 0.$$

The second term, which is the curl of the gradient of scalar is identically zero, then it follows, that always div.$\vec{B}$ = 0.

AMPERE'S LAW IN CURL FORM OR CURL $\vec{B}$ AND STOKES' THEOREM

The Ampere circuital law is a simple relation between magnetic vector $\vec{B}$ and current I. The law is easily applied to calculate $\vec{B}$ when sufficient symmetry exists. *If one is interested to apply the law the problems of a general nature without a high degree of symmetry, it is necessary to relate the magnetic vector* $\vec{B}$ *with the current density.* The situation here is rather like that div. $\vec{E} = \rho/t_o$, which relates the divergence of $\vec{E}$ with the charge density to specify the conditions at a point. Ampere's law can be written by using the definition of I and current density,

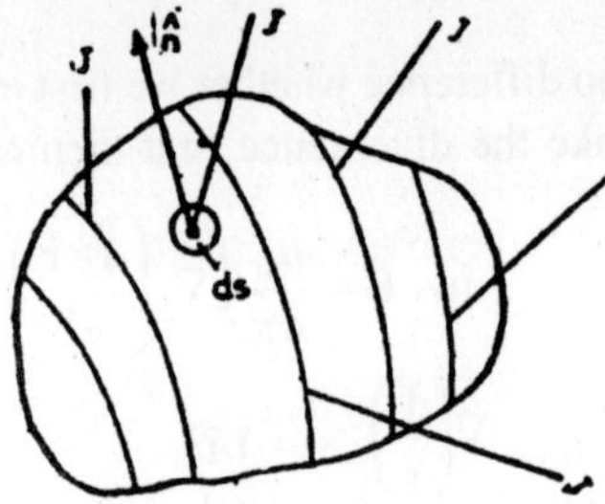

Fig. 5.18

$$\oint \vec{B}.d\vec{l} = \mu_o I$$

$$= \mu_o \int_8 \vec{J}.\hat{n}ds \quad ...(1)$$

where the area s is that specifying the region through which the current exists, $\hat{n}$ is a unit vector normal to elementary area ds in Fig. 3.10. Applying the Stokes' theorem,

$$\oint \vec{B}.d\vec{l} = \int_8 (\nabla \times \vec{B}).\hat{n}ds \quad ...(2)$$

Comparing Eqn. (1) and (2), we get,

$$\text{Curl } \vec{B} = \tilde{N} \times \vec{B} = \mu_o \vec{J} \quad ...(3)$$

Eqn. (3) is the point statement of Ampere's law corresponding to the integral statement, taking the two corresponding forms are,

$$\left.\begin{aligned} \oint \vec{B}.d\vec{l} &= \mu_o I \\ \nabla \times \vec{B} &= \mu_o J \end{aligned}\right\} \quad ...(4)$$

and

There expressions bear the same general relationship to each other that Gauss's law in electric field,

$$\oint \vec{E}.d\vec{s} = \frac{Q}{\varepsilon_o}$$

or, $\quad$ div. $\vec{E} = \rho/\varepsilon_o$

It is interesting to write down,

$$\left.\begin{aligned} &\text{div. } \vec{B} = \nabla . \vec{B} = 0 \\ \text{and} \quad &\text{Curl } \vec{B} = \nabla \times \vec{B} = \mu_0 J \\ \text{in electrostatic,} \quad & \\ &\text{div. } \vec{E} = \nabla . \vec{E} = \rho / \varepsilon_0 \\ \text{and} \quad &\text{Curl } \vec{E} = \nabla \times \vec{E} = 0 \end{aligned}\right\} \quad ...(5)$$

A Further Comment on Scalar Potential

We see from Eqn. (4), $\nabla \times \vec{B} = \mu_0 J$, that in a region containing no current flow curl $\vec{B} = 0$, but this does not hold within a conductor. In empty space where curl $\vec{B} = 0$, we can introduce as we have done a scalar potential but this cannot be done inside a conductor carrying a current. The scalar potential we introduced for a single loop of current was multiple-valued, that is as we integrated $\vec{B}$ around a contour enclosing the current, the potential increased by $\mu_0 I$, so that if we integrated n times around the contour we, should increase the potential by n $\mu_0 I$. In other words, for any single value of potential, all values obtained by adding or subtracting integral multiples of $\mu_0 I$ are equally legitimate. If now, we have many current loops, in our space, we can change the potential by integral multiples of μ_0 times of any one of the currents, by traversing suitable contours, with enough current loops, this means that we can obtain almost any value of the potential we desire, at a given point of space, if the current is distributed over the volume, so that we can enclose any amount by traversing a suitable contour within the volume, we can obtain any value of potential whatever at a given point of space. In this case the usefulness of the idea of potential breaks down completely. The integral of $\vec{B}$ is not independent of path, and there is no unique way of defining a scalar potential.

THE VECTOR POTENTIAL

It is clear from our discussion of the preceding section that a scalar potential is not useful for discussing a magnetic field, except in the special case where the current flows in a single loop. Mathematically, we have that curl $\vec{B}$ is not in general zero, so that the conditions for introducing a scalar potential do not exist. *There is, however, another quite different way of setting up a potential function, which is much more general, and equally useful. This is to set up what is called a 'vector potential'* .The fundamental characteristic of a potential is that it is a function that we differentiate to get the vector field. The reason why the potential is a useful device in electrostatics is that it is easy to compute it from charge distribution, by the solution of Poisson's equation. The vector potential solution for the magnetic field of currents has the same useful features. The existence of vector potential is based on div. $\vec{B}$ = 0, always satisfied by magnetic induction. By a well-known theorem of vector analysis, div. curl $\vec{A}$ = 0, where $\vec{A}$ is any arbitrary vector, that is, the divergence of any curl is zero. It seems reasonable to assume from this that we can write.

$$\vec{B} = \text{curl } \vec{A} \quad ...(1)$$

where $\vec{A}$ is a vector function of position, which is the vector potential. In other words, the vector $\vec{A}$ the curl of which is equal to the magnetic induction $\vec{B}$ is known as vector potential.

Eqn. (1) can be written as,

$$B_z\hat{k} = \left(\frac{\delta A_z}{\delta y} - \frac{\delta A_y}{\delta z}\right)\hat{i} + \left(\frac{\delta A_x}{\delta z} - \frac{\delta A_z}{\delta x}\right)\hat{j} + \left(\frac{\delta A_y}{\delta x} - \frac{\delta A_x}{\delta y}\right)$$

Now equating the components on both sides of the equation,

$$\left.\begin{aligned} \vec{B}_x &= (\nabla \times \vec{A})_x = \left(\frac{\delta A_z}{\delta y} - \frac{\delta A_y}{\delta z}\right) \\ \vec{B}_y &= (\nabla \times \vec{A})_y = \left(\frac{\delta A_x}{\delta_z} - \frac{\delta A_z}{\delta x}\right) \\ \vec{B}_z &= (\nabla \times \vec{A})_z = \left(\frac{\delta A_y}{\delta x} - \frac{\delta A_x}{\delta y}\right) \end{aligned}\right\} \quad ...(2)$$

where A_x, A_y, and A_z are the corresponding components of the vector potential A.

Divergence of $\vec{A}$

We shall consider in next section that div. $\vec{A}$ = 0.

Let us prove it. As div.$\vec{B}$ = 0, we have written,

$$\vec{B} = \nabla \times \vec{A}, \text{ it guarantees that,}$$

$$\nabla.B = \nabla.\,(\nabla \times \vec{A}) = 0$$

The field A is called the vector potential. In cases where $\nabla \times \vec{B} = 0$, we have $\vec{B} = -\nabla V_m$, where V_m is scalar potential. As we have discussed that the Scalar potential V_m was not completely specified by its definition. V_m is a multivalued function. If we found V_m for some problem, we can always find another potential that is equally good by adding a constant,

$$V_m' = V_m + C$$

The new potential V_m' gives the same magnetic field. So we can say that corresponding to a given value of magnetic field, there can be two or more scalar potentials differing by addition of a constant C, since gradient ∇ C is zero we have V'_m and V_m represent the same phenomena i.e, the same magnetic field. Similarly, we can have different vector potentials $\vec{A}$ which give the same magnetic field. Again, because $\vec{B}$ is obtained from $\vec{A}$ by differentiating as given in Eqn. (2) adding a constant to A does not change anything physical. But there is even more latitude for A. We can add to A any field which is the gradient of some scalar field, without changing to physical phenomena.

Suppose we have a vector potential A that gives correctly the magnetic field $\vec{B}$ and the other new vector potential A' will give the same field $\vec{B}$ if they have the same curl *i.e.*,

$$\vec{B} = \nabla \times \vec{A} = \nabla \times \vec{A}'.$$

Therefore, $\nabla \times \vec{A} - \nabla \times \vec{A} = \nabla \times (\vec{A}' - \vec{A}) = 0$

But if the curl of a vector is zero it must be the gradient of some scalar field, say ϕ. So,

$$A' - A = \nabla \phi$$

$$A' = A + \nabla \phi \qquad ...(3)$$

will be an equally satisfactory vector potential, leading to the same field $\vec{B}$. It is usually convenient to vary $\vec{A}$ by imposing some condition on it (in the same way that we found it convenient to choose the potential V_m zero at large distances.) We can, for instance, restrict $\vec{A}$ by choosing arbitrarily value of divergence of $\vec{A}$, we can always do without affecting $\vec{B}$. This is because although A' and A have the same curl, and give the same $\vec{B}$, they do not need to have the same divergence. In fact,

$$\nabla . \vec{A}' = \nabla . \vec{A} + \nabla^2 \phi$$

And by a suitable Choice of ϕ, we can make $\nabla . \vec{A}'$ any thing we wish.

What should be the choice of $\nabla . \vec{A}$? The choice should be made to get the greatest mathematical convenience and will depend on the individual problem. For magnetostatics, we will make the simple choice,

$$\nabla . \vec{A} = 0$$

Then the complete definition of $\vec{A}$ is for magnetostatics.

$$\nabla \times \vec{A} = \vec{B}$$

and $$\nabla . \vec{A} = 0 \quad ...(4)$$

THE VECTOR POTENTIALS OF KNOWN CURRENTS

As we know that the magnetic field vector $\vec{B}$ is determined by the currents, the vector potential $\vec{A}$ is related to magnetic field B by the relation,

$$\vec{B} = \text{curl } \vec{A},$$

So $\vec{A}$ must also be determined by the currents. We want to find $\vec{A}$ in terms of the currents. We begin with the basic equation,

$$\nabla \times \vec{B} = \mu_o \vec{J},.$$

Substituting $\vec{B}$ in terms of vector $\vec{A}$, the equation reduces, to

$$\nabla \times \nabla \times \vec{A} = \mu_o \vec{J}$$

using the vector identity.

$$\nabla \times \nabla \times A = \nabla (\nabla . \vec{A}) - \nabla^2 A = \mu_o J$$

Now since $\nabla . A = 0$, therefore,

$$\nabla^2 \vec{A} = \mu_o \vec{J}$$

$$\left.\begin{aligned} &\nabla^2 \vec{A} = \mu_o \vec{J} \\ &\text{or in terms of components,} \\ &\nabla^2 A_x = -\mu_o J_x \\ &\nabla^2 A_y = -\mu_o J_y \\ &\text{and} \\ &\nabla^2 A_z = -\mu_o J_z \end{aligned}\right\} \quad ...(1)$$

Each of the equation given in eq. (1) is mathematically identical to the equation of electrostatics *i.e.*, Poisson's equation,

$$\nabla^2 \vec{A} = -\rho/\varepsilon_o$$

Now we know that the solution of Poisson's equation gives the value of V(x, y,z), hence a very simple process for calculating electrostatics field of a charge distribution,

$$V(x_1, y_1, z_1) = \frac{1}{4\pi\varepsilon_o} \int_V \frac{\rho(x_2, y_2, z_2)}{r_{12}} dv_2$$

where r^{12} is the separation between the points (x_2, y_2, z_2) and (x_1, y_1, z_1). So we see immediately from an analogy the general solution for equation,

$$\nabla^2 A_x = -\mu_o J_x$$

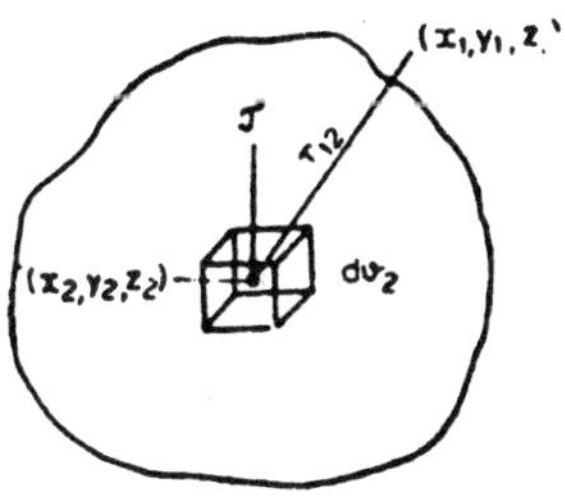

Fig. 5.19

As, $$A_x = \frac{\mu_o}{4\pi} \int_V \frac{J_x(x_2, y_2, z_2)}{r_{12}} dv_2$$

i.e., simply replacing ρ/ε_o by $\mu_o J_x$.

The dv_2 and r_{12} are shown in Fig. 4.26. Similarly we have for A_y and A_z. And when three solutions are combined, we get the solution of

$\nabla^2 A = \mu_o \vec{J}$, *i.e.*,

$$A(x_1, y_1, z_1) = \frac{\mu_o}{4\pi} \int_V \frac{J_x(x_2, y_2, z_2)}{r_{12}} dv_2 \qquad ...(2)$$

One can verify by direct differentiation that this integral satisfies condition $\nabla \cdot \vec{A} = 0$, so long as $\nabla \cdot \vec{J} = 0$, which must happen for steady currents. So by using Eqn. (2), we can find vector potential $\vec{A}$ due to the current density $\vec{J}$.

If the current is flowing in a wire having diameter very small as compared with the dimension, equation (2) can be modified for a thin wire of area of cross-section S and length dl, dv = Sdl,

hence, $$\text{J}dv = \text{JS}dl$$

as current density is defined as current per unit area, dS = I, hence

$$\text{J}dv = \text{I}dl$$

Putting in equation 2, we have,

$$A = \frac{\mu_o}{4\pi}\oint\frac{\text{Idl}}{r_{12}} \qquad ...(3)$$

where r_{12} is the distance of dl to the point at which we are calculating potential and the integral is taken around a whole curcuit.

APPLICATIONS OF VECTOR POTENTIALS

As, an example, we shall calculate the magnetic field by using vector potential method.

1. The B of Straight Wire

Let us consider a long straight wire of radius a carrying a steady current I which is uniformly distributed throughout its cross-section and the wire is placed in coordinate system as shown in

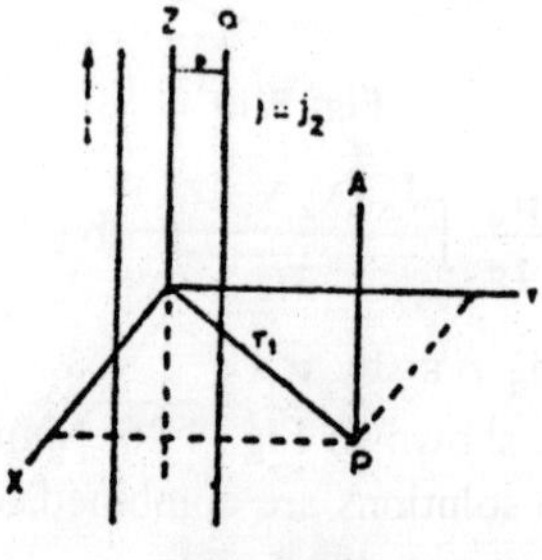

Fig. 5.20

Fig. 5.20 along the z-axis. The current density vector $\vec{J}$ will have, in this case, only z-component. It s magnitude is given by,

$$J_z = I/\pi a^2$$

inside the wire and is zero outside. Since $J_x = 0$ and $J_y = 0$, we immediately have,

$$A_x = 0,\ A_y = 0$$

To get A_z we can use our solution for the electrostatics potential V of a wire with a uniform charge density. For points outside, an infinite charge cylinder, the electrostatic potential is,

$$V = \frac{\lambda}{2\pi\varepsilon_o} \log r'$$

where $r' = \sqrt{x^2 + y^2}$ and λ is the charge per unit length and is equal to the charge contained in a unit length of the cylinder, *i.e.*,

$$\lambda = \pi a^2 \rho$$

therefore, $$V = \frac{I}{2\pi\varepsilon_o} \pi a^2 \rho \log r'$$

The corresponding vector potential Az is obtained by replacing ρ/ε_o by $\mu_o J_z$, *i.e.*,

$$A_z = -\frac{I}{2\pi} \pi a^2 \mu_o J_z \log r'$$

putting $I = \pi a^2 J_z$,

$$A_z = -\frac{\mu_o I}{2\pi} \log r'$$

Now, we can find B by using Eqn. (1),

$$B_x = \left(\frac{\delta A_z}{\delta_y} - \frac{\delta A_y}{\delta_z}\right)$$

Putting A_z given above and noting that $A_y = 0$, we get,

$$B_x = -\frac{\mu_o I}{2\pi} \frac{\delta}{\delta y}(\log r'),\ r' = \sqrt{x^2 + y^2}$$

After simplification,

$$B_x = -\frac{\mu_o I}{2\pi} \frac{y}{r'^2}$$

Similarly, $$B_y = \left(\frac{\delta A_x}{\delta_z} - \frac{\delta A_z}{\delta_x}\right)$$

as $A_x = 0$, and substituting A_z

$$= \frac{\mu_o I}{2\pi} \frac{\delta}{\delta x}(\log r') = \frac{\mu_o I}{2\pi} \frac{x}{r'^2}$$

and, $$B_z = \left(\frac{\delta A_y}{\delta x} - \frac{\delta A_x}{\delta y}\right) = 0, \text{ since } A_x = A_y = 0.$$

The magnitude of the magnetic field B is given by,

$$B = (B_x^{\ 2} + B_y^{\ 2} + b_z^{\ 2})^{1/2}$$

Substituting the values obtained above, we get

$$B = \frac{\mu_0 I}{2\pi r'^2}(x^2 + y^2)^{1/2}$$

$$= \frac{\mu_0 I}{2\pi r'} \qquad ...(2)$$

This is the same result as obtained earlier by Biot-Savart law and Ampere's circuital law.

2. The Field of a Small Current Loop-The Magnetic Dipole

Let us use the vector potential method to find the magnetic field of a small loop of current. By small loop, we mean simply that we are interested in the fields only at large distances compared with the size of the loop. Such a loop produces magnetic field like the electric field from an electrical dipole. It will be shown below that any small loop is a magnetic dipole.

Consider a rectangular loop and choose out coordinates as shown in Fig. 5.21(a). There are no currents in the z-direction so A_z is zero. There are currents in the x-direction on the two-sides of length a. In each leg, the current density (and current) is uniform. So the solution for A_x is just like the electrostatic potential from two charged rods s shown in Fig. 5.21(b). Since the rods

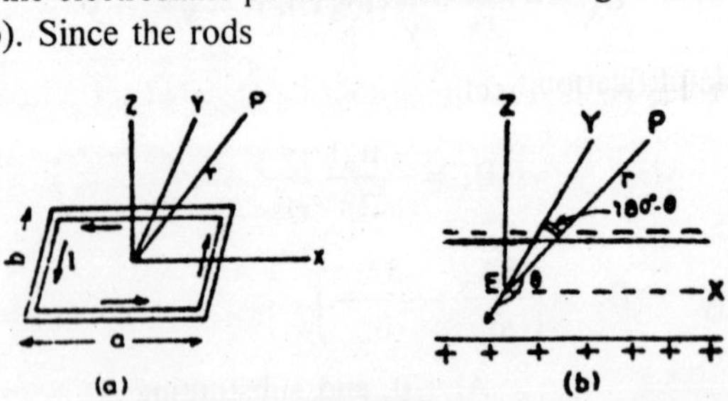

Fig. 5.21

have opposite charges, the electric potential at large distances would be just the dipole potential at the point P, the potential would be,

$$V = \frac{I}{4\pi\varepsilon_0}\frac{\vec{p}.\hat{r}}{r^2}$$

where P is the dipole moment of the charge distribution.

The dipole moment in the present case is the total charge on one rod times the separation between them.

$$P = \lambda\ a\ b$$

where λ is the linear density of charge on the rods.

If α is the are of cross-section of the wire, ρ is the charge per-unit volume, them

$$\lambda - \text{volume of unit length of wire} \times \rho$$

$$= \alpha\ \rho$$

hence,

$$P = \alpha\ \rho\ ab$$

and

$$V = \frac{1}{4\pi\varepsilon_0}\frac{P\cos\theta}{r^2}$$

where θ is the angle between the direction of $\vec{p}$ and $\vec{r}$ dirction of $\vec{p}$ is same as $\vec{E}$ with negative y-direction,

$$\cos(180 - \theta) = y/r$$

$$\cos\theta = -y/r$$

therefore,
$$V = -\frac{1}{4\pi\varepsilon_0}\frac{P}{r^2}\frac{y}{r}$$

putting the value of P given above,

$$V = -\frac{1}{4\pi\varepsilon_0}\frac{\alpha\rho ab}{r^2}\frac{y}{r}$$

From this A_x is directly written by replacement of ρ/ε_0 by $\mu_0 J$ *i.e.*,

$$A_x = -\frac{1}{4\pi}\frac{\mu_0 J\alpha\, ab}{r^2}\frac{y}{r}$$

$$J\ \alpha = \text{current } I$$

hence,
$$A_x = -\frac{\mu_0 I ab}{4\pi}\frac{y}{r^3} \quad ...(3)$$

Similarly, A_y can be calculated from the electric potential due to two oppositely charged rod lying parallel to y-axis separated by distance α.

$$V = \frac{1}{4\pi\varepsilon_o} \frac{P\cos\theta}{r^2}$$

Since $\cos\theta = x/r$,

$$V = \frac{I}{4\pi \in b} \frac{\rho\alpha\, ab}{r^2} \frac{x}{r}$$

hence A_y is obtained by replacing ρ/ε_o by $\mu_o J$

$$A = \frac{\mu_o I\, ab}{4\pi} \frac{x}{r^3} \qquad ...(4)$$

Therefore, the vector potential

$$A = (A_x{}^2 + A_y{}^2 + A_z{}^2)^{1/2}.$$

Putting A_x and A_y obtained above and also noting that $Az = 0$,

we get, $$A = \frac{\mu_o I\, ab}{4\pi} \frac{\left(x^2+y^2\right)^{1/2}}{r^3}$$

but $$r^2 = x^2 + y^2 + z^2$$

hence $$x^2 + y^2 = r^2 - z^2$$

or $$(x^2 + y^2)^{1/2} = r\sqrt{1} - x^2/r^2 \; z/r = \cos\theta_z$$

hence $$A = \frac{\mu_o I\, ab}{4\pi} \frac{1}{r^2} \sqrt{1-\cos^2\theta_z}$$

where θ_z is the angle which r makes with z-axis,

$$A = \frac{\mu_o}{4\pi} I\, ab \frac{\sin\theta_z}{r^2}$$

The quantity Iab which determines the strength of A and is product of area of the loop and current. This product is known as m, the magnetic dipole moment. Hence,

$$A = \frac{\mu_o}{4\pi} \frac{m\sin\theta_z}{r^2}$$

$$= \frac{\mu_o}{4\pi} \frac{\vec{m}\times\hat{r}}{r^2} \qquad ...(5)$$

This will be the same at every point on a circle around the dipole. $\overline{A}$ is perpendicular to the plane containing the dipole and point of observation .

The magnetic field $\vec{B}$ can be calculated from Eqn. (6), as,

$$B_x = \left(\frac{\partial A_z}{\partial y} - \frac{\partial A_y}{\partial z}\right)$$

$$= -\frac{\delta A_y}{\delta_z}, \text{ since } A_z = 0.$$

Substituting A_y from Eq. (7), B_x reduces as given below.

$$B_x = -\frac{\partial}{\partial_z}\left(\frac{\mu_0 m}{4\pi} \cdot \frac{x}{r^3}\right) \text{ where } m = \text{Iab and } r = (x^2 + y^2 + z^2)^{1/2}$$

Substituting this, B_x reduces to

$$B_x = \frac{\mu_0 m}{4\pi} \cdot \frac{3xz}{r^5} \qquad \text{....(6)}$$

Similarly,

$$B_y = \frac{\mu_0 m}{4\pi} \cdot \frac{3yz}{r^5}$$

The value B_z is given in Eqn. (7), *i.e.*,

$$B_z = \frac{\partial A_y}{\partial x} - \frac{\partial A_x}{\partial y}$$

Substituting the values from Eqn. (7) and Eqn. (7a),

$$B_z = \frac{\mu_0 m}{4\pi}\left[\frac{\partial}{\partial x}\left(\frac{x}{r^2}\right) + \frac{\partial}{\partial y}\left(\frac{y}{r^3}\right)\right]$$

$$= \frac{\mu_0 m}{4\pi}\left[\frac{2}{r^3} - \frac{3}{r^5}\left(x^2 + y^2\right)\right]$$

Since, $r^2 = x^2 + y^2 + z^2$

after substitution, we get

$$B_z = -\frac{\mu_0 m}{4\pi}\left(\frac{1}{r^3} - \frac{3z^2}{r^5}\right) \qquad \text{...(7)}$$

therefore, the magnitude of B,

$$B = [B_x^2 + B_y^2 + B_z^2]^{1/2}$$

On substituting the values obtained above, we get,

$$B = \frac{\mu_0 m}{4\pi r^3}\sqrt{1 = 4\cos^2\theta} \qquad \text{...(8)}$$

This expression is same as that of the electric dipole oriented along the z-axis. The current loop is, therefore, called a magnetic dipole.

3. Change in Magnetic Field at a Current Sheet

Consider a thin sheet of copper of uniform thickness t located in the xz plane in which a current flows with constant density and direction everywhere within the metal. Let the current flows is the x-direction. Fig. 5.22 shows a portion of the sheet of infinite extent. If the current density inside the metal is j in amp/metre^2 then every metre of height *i.e.*, of sheet in z-direction have an area t metre^2 and will include a ribbon of current amounting to jt ampere. We call this the surface current density or simply sheet current density and represented by j_8, j_8 is a useful quantity and determines the change in the magnetic field from one side of the sheet to the other.

The total magnetic field is the sum of the two fields, one due to sheet and the other from some external source present, there, Consider a rectangular path abcd, ab side is in front of the surface and dc behind with the short sides bc and da piercing the sheet. Let B_z + sheet and B_z – be that just behind it the line integral of B around abcd, is given by,

$$\int_{abcd} \vec{B}.d\vec{l} = \int_{ab} \vec{B}.d\vec{l} + \int_{bc} \vec{B}.d\vec{l} + \int_{cd} \vec{B}.d\vec{l} + \int_{dc} \vec{B}.d\vec{l}$$

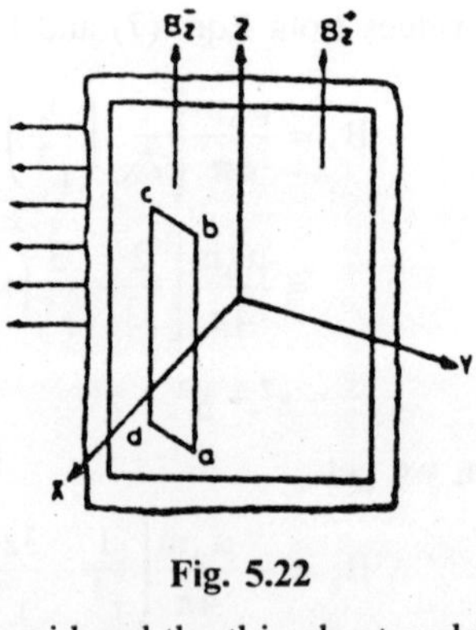

Fig. 5.22

Since we have considered the thin sheet and the point just near the sheet, hence bc and da sides are much shorter than ab and cd. Hence their contribution is neglected. Therefore,

$$\int_{abcd} \vec{B}.d\vec{l} = \int_{ab} \vec{B}.d\vec{l} + \int_{cd} \vec{B}.d\vec{l}$$

$$= (B_z^+ - B_z^-).l \quad ab = cd = l$$

Fields in front of the sheet and just behind it are just opposite to each other. Now using the Ampere's law

$$\int_{abcd} \vec{B}.d\vec{l} = \mu_o \times \text{current enclosed by rectangle}$$

$$= \mu_o \, j_s l$$

or, $$(B_z^+ - B_z^{-1})\, l = \mu_o \, j_s l$$

or, $$B_z^+ - B_z^- = \mu_o \, j_s \qquad ...(9)$$

A current sheet of density j_s increases the component of $\vec{B}$ which is parallel to the surface and perpendicular to j_s. This increase is the same as the change in electric field at a sheet of charge. If the sheet is the only current source (the other external source is contributing). Then, of course, the field is symmetrical about the sheet. We have,

$$B_z + = \frac{1}{2}\mu_o j_8$$

and $$B_z - = -\frac{1}{2}\mu_o j_8$$

Suppose there are two sheets carrying equal and opposite surface currents as shown in Fig. 5.23, with on other sources around. The direction of current flow is perpendicular to the plane of the paper, out on the left and in on the right. The field between the sheet is and outside the sheets is zero

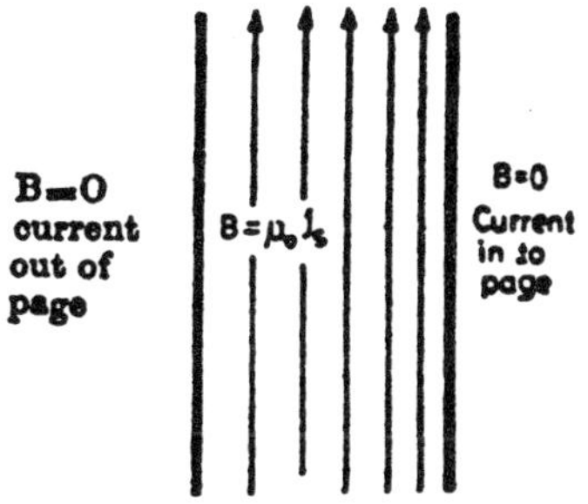

Fig. 5.23

$$\mu_o j_z = \left(\frac{1}{2}\mu_o j_8 + \frac{1}{2}\mu_o j_8\right)$$

$$\frac{1}{2}\mu_o j_8 + \frac{1}{2}\mu_o j_8 = 0$$

ASTON'S MASS SPECTROGRAPH

It is an improvement on Thomson method to increase the accuracy by modifying the principle of operation.

Principle

The main modification introduced is that the electric and magnetic fields are not applied simultaneously and in parallel direction as Thomson did, but successively, first the electric field and then the magnetic field.

Their directions being at right angles to each other. Therefore, the electric field produces a dispersion of the positive rays with respect to the velocity, while the magnetic field brings to a focus the dispersed rays.

Apparatus and its Working

A narrow beam of positive ions is made to pass through two parallel narrow slits A and B (0.01 mm in radius and 20 cm. apart) and a very fine beam is obtained. This beam is subjected to a uniform electric field intensity maintained between the plates C and D. The beam as a whole is deflected towards the negative electrode D. The beam is also dispersed through a certain angle because the beam contains ions of different velocities. The fast moving ions remain for a shorter time under the influence of electric field will be deflected less than the slower ones taking a longer time to traverse the field. Thus the beam is spread out into an electric spectrum according to different velocities. The divergent beam limited by slit S next enters the magnetic field $B \perp E$ and the beam is bent in a direction opposite to that caused by E. Again the slowly moving ions are deflected more than the faster ions. Thus the beam can be focused at some point on photographic plate, if S is the linear displacement of the ions and l the length of their path in the electric field, then from Eq. (1),

$$S = \frac{eEl^2}{2mv^2}$$

(Force eE = ma, a = eE/m, initial velocity is zero

$$\left. S = \frac{1}{2}at^2,\ t = \frac{l}{v},\ S = \frac{1}{2}\frac{eE}{m}\frac{l^2}{v^2} \right)$$

Similarly if S' is linear displacement and l' the length of path of the ion in the magnetic field

$$S' = \frac{1}{2}\frac{eB}{mv}l'^2$$

$$\left(\text{Force} = evB = ma', \quad a' = \frac{evB}{m}, S' = \frac{1}{2}a' t'^2, t' = \frac{l'}{v} \right.$$

$$\left. S' = \frac{1}{2} \frac{evB}{m} \frac{l'^2}{v^2} = \frac{1}{2} \frac{eB}{mv} l'^2 \right)$$

Let θ and ϕ be the mean angles of deviation in the electric and magnetic fields respectively,

$$\theta = S/l \quad \text{and} \quad \phi = S'/l'$$

$$\theta = \frac{eEl}{2mv^2} \qquad \phi = \frac{eBl'}{2mv}$$

or $$\theta = K_e \frac{e}{mv^2} \qquad ...(2)$$

and $$\phi = K_m \frac{e}{mv} \qquad ...(3)$$

where K_e and K_m are constants, depending on the strength and distribution of electric and magnetic fields. Let dθ be the dispersion angle due to electric field and dϕ the convergence angle due to the magnetic field. Differentiating equations (2) and (3), we get,

$$d\theta = -2K_e \frac{e}{mv^3} dv = -2\theta \frac{dv}{v}$$

or $$\frac{d\theta}{\theta} = -2\frac{dv}{v} \qquad ...(4)$$

Similarly, $$d\phi = -K_m \frac{e}{mv^2} dv$$

$$= -\phi \frac{dv}{v}$$

or $$\frac{d\phi}{\phi} = -\frac{dv}{v} \qquad ...(5)$$

Comparing Eqs. (4) and (5), we get

$$\frac{d\theta}{\theta} = 2\frac{d\phi}{\phi} \qquad ...(6)$$

Let OR = a, and RL = b

The dispersion of the selected group at R is adθ. In the absence of magnetic field the dispersion produced in the beam for a distance a + b

$$= (a + b)\, d\theta$$

As this dispersion (or divergence) is annulled by the magnetic field acting at R so that the group comes to focus at L, distance b from R the condition of focusing is given by,

$$(a + b)\, d\theta = b d\phi$$

$$\frac{a+b}{b} = \frac{d\phi}{d\theta} \qquad ...(7)$$

$$\frac{a+b}{b} = \frac{\phi}{2\theta}, \text{ from eqn. (6)}$$

$$(a + b)\, 2\theta = b\phi$$

$$b\,(\phi - 2\theta) = 2a\theta \qquad ...(8)$$

This is the condition for focusing. If plate P is inclined at an angle α to the incident beam AB and RN be the perpendicular on the line joining O and plate P, from triangle RNO

$$RN = a \sin(\alpha + \theta) \qquad ...(9)$$

and from triangle RNL,

$$RN = b \sin RLN$$

$$RN = b \sin(\phi - \alpha - \theta) \qquad ...(10)$$

Equating eqns. (9) and (10),

$$a \sin(\alpha + \theta) = b \sin(\phi - \alpha - \theta)$$

for small angle,

$$a(\alpha + \theta) = b(\phi - \alpha - \theta) \qquad ...(11)$$

Comparison of eqns. (10) and (11) shows that two equations are identical if $\alpha = \theta$. *Thus the condition for focussing is that photographic plate must be placed at an angle θ with the direction of incident positive ray.*

Separation of Isotopes by Analysis of Traces

Substance of known atomic masses are mixed up with elements whose isotopes are to be determined and traces are obtained on photographic plate. The traces of the known atomic mass serve as reference, thus the traces of unknown isotopes are known by measuring the distance from the reference point.

Advantages of Aston's Mass Spectrograph Over Thomson Spectrograph

The modified technique of Aston have the following advantages over Thomson method:

1. The dispersion of the ionic beam according to different velocities of ions is greater than in Thomson mass spectrograph.
2. All the ions having the same e.m are focussed at one pint, instead of spreading them into a parabola thus securing greater intensity for photography. Hence a fine slit can be used to obtain sharp lines to be obtained on the photographic plate, thereby securing greater sensitivity. The accuracy was 1 in 1,000.

MAGNETIC FIELD OF A MOVING POINT CHARGE (LAPLACE RULE)

It was found experimentally that the magnetic induction resulting from a charge q moving with velocity $\vec{v}$ at a distance r away from the charge where r is a vector pointing from the charge to the point where the field is being found, is related by,

$$B \propto \frac{qv \sin\theta}{r^2}$$

where B is the magnitude of the induction and θ is the angle between $\vec{v}$ and $\vec{r}$ as shown in Fig. 5.24.

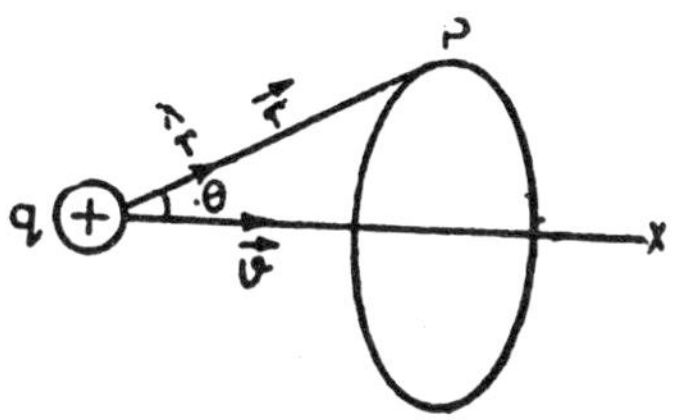

Fig. 5.24

The above equation can be written in terms of equality by inserting a constant $m_0/4\pi$. Therefore,

$$B = \frac{\mu_0}{4\pi}\frac{qv \sin\theta}{r^2}$$

or $$\vec{B} = \frac{\mu_0}{4\pi} q\left(\frac{\vec{v}\times\vec{r}}{r^2}\right) \qquad ...(1)$$

or $$\vec{B} = \frac{\mu_0}{4\pi} q\left(\frac{\vec{v}\times\vec{r}}{r^3}\right) \qquad ...(1a)$$

Where $\vec{r}$ is unit vector along r,

μ_0 is called permeability of free space.

We have chosen the permeability constant $\frac{\mu_o}{4\pi}$ as for Coulomb's law, we introduced $\frac{1}{4\pi\varepsilon_o}$ and for the same reason *i.e.*, in rationalised MKS system the factors π should occur in problems having spherical, cylindrical or circular symmetry and should not occur when these symmetries do not exist. In fact, certain equations that are derived from equation (1) but are used more often than it is, will be simpler in form if we introduce $\frac{\mu_o}{4\pi}$. *We shall find the physical significance of the number just later.*

The direction of $\vec{B}$ is determined by the rules of vector product. From equation (1), $\vec{B}$ is perpendicular to $\vec{v}$ and $\vec{r}$, *i.e.*, in the direction normal to the plane containing $\vec{v}$ and $\vec{r}$ which is plane of the paper.

SOLVED EXAMPLES

Example 1:

If a stream of electrons, each of mass m and charge e and velocity 3×10^6 metre/sec. is deflected by 2 mm in traversing a distance of 10cm through an electrostatics field of 18 V/cm. perpendicular to their path, find e/m in couls/kg.

Solution:

$$E = 18\,\frac{\text{Volts}}{\text{cm}} = \frac{18}{10^{-2}}\,\frac{\text{Volts}}{\text{metre}}$$

$$eE = ma$$

$$y = \frac{1}{2}at^2 = \frac{1}{2}\left(\frac{eE}{m}\right)\left(\frac{1}{v}\right)2$$

$$2\times10^{-3} = \frac{1}{2}\left(\frac{18}{10^{-2}}\cdot\frac{e}{m}\right)\left(\frac{10\times10^{-2}}{3\times10^{6}}\right)^2$$

or

$$\frac{e}{m} = \frac{2\times10^{-3}\times2\times10^{-2}\times9\times10^{12}}{18\times100\times10^{-4}}$$

$$= 2 \times 10^{11} \text{ couls. kg.}$$

Example 2:

For a proton initially travelling with velocity $vx = 5 \times 10^8$ cm/sec. Calculate the following:

(1) *Force experienced in a magnetic field* $\vec{B} = 2000\ \hat{j} + 4000\ \hat{k}$ *gauss*

(2) *Acceleration experienced in an electric field E,*

$$\vec{E} = 200\hat{i} + 100\hat{j} \text{ volts/cm}$$

(3) *Transverse deflection in travelling a length* $L_x = 10$ *cm in an electric field* $E_y = 200$ *volts/cm.*

Solution:

(1) Force in magnetic field

$$\vec{F} = q(\vec{v}\times\vec{B})$$

$$\vec{v}_x = 5 \times 10^8 \text{ cm/sec} = 5 \times 10^6 \text{ metre/sec}$$

$$\vec{B} = 2000\ \hat{j} + 4000\ \hat{k}$$

$$\vec{B} = 0.2\ \hat{j} + 0.4\ \hat{k} \text{ weber/m}^2$$

Since, $\dfrac{1 \text{ Weber}}{\text{metre}^2} = 10^4$ gauss

hence, $\vec{F} = 1.6 \times 10^{-19} [(5 \times 10^6\ \hat{i}) \times (0.2\ \hat{j} + 0.4\ \hat{k})]$

$$= 1.6 \times 10^{-19} \times 5 \times 10^6 \times 2\ [0.1\hat{k} - 0.2\ \hat{j}]$$

$$= 1.6 \times 10^{-13}\ [\ \hat{k} - 2\hat{j}] \text{ Newtons}$$

Magnitude of the force F = 1.6×10^{-13} Ö5 = 3.57×10^{-13} Nt.

Its direction cosines

$$\left((0, -\frac{2}{\sqrt{5}}, \frac{1}{\sqrt{5}}\right)$$

(2) In electric field,

$$ma = qE$$

$$a = \frac{qE}{m} = \frac{1.6\times10^{-19}}{1.67\times10^{-27}}\left(\frac{200\hat{i}+100\hat{j}}{10^{-2}}\right)$$

$$= \frac{1.6}{1.67}10^{12}\ (2\hat{i}+\hat{j})$$

$$= \frac{1.6}{1.67}10^{12}\sqrt{5}$$

$$= 2.14 \times 1012 \text{ met/sec}^2$$

(3) Transverse deflection,

$$Y = \frac{1}{2}at^2$$

$$= \frac{1}{2}\left(\frac{gE_y}{m}\right)\left(\frac{1}{v_x}\right)^2,$$

$$\text{Since } E_y = 200\ \frac{\text{volt}}{\text{cm}} = \frac{200}{10^{-2}}\ \frac{\text{volt}}{\text{metre}}$$

$$Y = \frac{1}{2} \times \frac{1.6\times10^{-19}\times200}{1.67\times10^{-27}\times10^{-2}} \times \left(\frac{10\times10^{-2}}{5\times10^{-6}}\right)^2 \text{ metre}$$

$$= 0.0383 \times 10^{-2} \text{ metre}$$

$$= 0.0383 \text{ cm}$$

Example 3:

An electron of energy 20 eV moving in a direction perpendicular to a magnetic field B = 10^2 weber/m^2 describes a circle or radius R. Calculate the value of R.

Solution:

Kinetic energy $1/2\ mv^2 = 20$ eV

$$= 20 \times 1.6 \times 10^{-19} \text{ joules}$$

hence, $$v = \left(\frac{2\times20\times1.6\times10^{-19}}{m}\right)^{1/2} \text{ m / sec}$$

The radius of the circular path is given by equation (1),

$$R = \frac{mv}{qB}$$

$$= \frac{m\times\left(64\times10^{-19}\right)^{1/2}}{\sqrt{m\times q\times B}}$$

$$= \frac{\left(9\times10^{-31}\times64\times10^{-19}\right)^{1/2}}{1.6\times10^{-19}\times10^2}$$

$$= 15.0 \times 10^{-8} \text{ metre}$$

Example 4:

An electron of velocity $\vec{v} = (2\hat{i} + 3\hat{j})\ 10^8$ cm/sec. enters a region of uniform magnetic field, B = 500 $\hat{i}$ gauss, show that its path becomes helical.

(1) In what direction does the axis of the helix lie ?

(2) Calculate the radius of helix

(3) Calculate the number of rotations as the electron advances 10 cm along the axis of the helix.

Solution:

$\vec{v} = (2\hat{i} + 3\hat{j})\ 10^8$ cm/sec. and

$\vec{B} = 500\hat{i}$ gauss

The component of velocity $\vec{v}_1 = 2\hat{i} \times 10^8$ cm/sec. is along $\vec{B}$ and electron moves along x-axis and due to the component $\vec{v}_2 = 2\hat{j} \times 10^8$ cm/sec. which is perpendicular path.

(1) The axis of helix lines along B *i.e.*, x-axis.

(2) Radius of helix R by equation (1)

$$R = \frac{mv_2}{qB} + \frac{9\times10^{-31}\times3\times10^6}{1.6\times10^{-19}\times500\times10^{-4}}$$

(1 Weber/metre^2 = 10^4 gauss)

$$R = 3.37 \times 10\text{–}4 \text{ metre}$$

$$= 0.0337 \times 10\text{–}2 \text{ metre}$$

$$= 0.0337 \text{ cm}$$

(2) Distance travelled in one rotation = path of the helix by equation (2),

$$\frac{2\pi m\, v_1}{qB} = \frac{2\pi\times9\times10^{-31}\times2\times10^6}{1.6\times10^{-19}\times500\times10^{-4}}$$

Number of rotations made by the electron as it advances through a distance of 10 cm

$$= \frac{10\times10^{-2}}{\text{pitch of helix}}$$

$$= \frac{10\times10^{-2}\times1.6\times10^{-19}\times500\times10^{-4}}{2\pi\times9\times10^{-31}\times2\times10^6}$$

$$= 70.74 = 71.$$

Example 5:

A long straight wire carries a current of 20 amperes. An electron is travelling at 10^7 m/sec, is 2.0 cm from the wire, what force acts on *the electron if its motion is directed (1) towards the wire, (2) parallel to the wire and, (3) at right angles to the direction given in (1) and (2)*

Solution:

The magnitude of B produced by wire carrying currents is,

$$B = \frac{\mu_o I}{2\pi r} \frac{4\pi \times 10^{-7} \times 20}{2\pi \times 0.02/100}$$

$$= 2 \times 10^{-4} \text{ weber/metre}^2$$

Force on the electron $\vec{F} = q(\vec{v} \times \vec{B})$

(1) When electron moves towards the wire, v is perpendicular to the B due to wire so,

$$F = qvB = 1.6 \times 10^{-19} \times 10^7 \times 2 \times 10^{-4}$$

$$= 3.2 \times 10^{-16} \text{ Newtons.}$$

F is parallel to the current.

(2) When electron moves parallel to the wire, its velocity v will again be perpendicular to the magnetic field B. So the magnitude will be = 3.2×10^{-16} Newtons.

(3) When the electron is moving at right angle to the direction given above then v and B will be parallel or anti-parallel.

Hence, $\vec{F} = q(\vec{v} \times \vec{B})$ will be zero

or $F = 0$

Example 6:

A solenoid is 1 metre long an d3 cm in mean diameter. It has 5 layers of windings of 850 turns each and carries a current of 5.00 amp. Calculate B at its centre.

Solution:

$$B = \mu_o I_o n$$

$$= 4\pi \times 10^{-7} \times 5 \times 5 \times 850$$

$$= 2.7 \times 10^{-2} \text{ weber/metre}^2$$

Example 7:

The electron circulates around the nucleus in a path of radius 5.1 × 10^{-11} m at a frequency of 6.8 × 10^{15} revolutions/second. Calculate B at the centre and the magnetic dipole moment.

Solution:

$$\text{Current I} = \text{Charge flow/sec} = ev$$
$$= 1.6 \times 10^{-19} \times 6.8 \times 10^{15}$$
$$= 1.1 \times 10^{-3} \text{ amp.}$$

B at the centre of the orbit is given by putting x = 0 in Eq. (1).

$$i.e., \; B = \frac{\mu_0 IR^2}{2(R^2 + x^2)^{3/2}} = \frac{\mu_0 I}{2R} = \frac{4 \times 10^{-7} \times 1.1 \times 10^{-3}}{2 \times 5.1 \times 10^{-11}}$$
$$= 14 \text{ weber/metre}^2$$

Magnetic dipole moment from Eq. (2) of Chapter 9,

$$m \; NIA = 1 \times 1.1 \times 10^{-3} \times \pi \times (5.1 \times 10^{-11})^2$$
$$= 9 \times 10^{-24} \text{ amp.metre}^2.$$

Example 8:

Find an expression for the magnetic induction at the centre of a circular current loop.

Solution:

Magnetic field at the centre of a circular current loop.

Consider a circular current loop of radius R carrying current I. Let us first calculate the magnetic from $\vec{B}$ due to small element $d\vec{l}$.

$$d\vec{l} = Rd\,\theta\;\hat{\theta}_o$$

where is the a unit vector in the θ direction. Note that since $d\vec{l}$ is small means dθ is small. The unit vector.

$\hat{r}$ and $\hat{\theta}_o$ are perpendicular to each other at all points along the path. The, from equation (2), the field due to this element at the centre,

$$d\vec{B} = \frac{\mu_o}{4\pi}\frac{I(d\vec{l} \times \hat{r})}{R^2}$$
$$= \frac{\mu_0 I}{4\pi}\left(\frac{Rd\theta\;\hat{\theta}_o \times \hat{r})}{R^2}\right)$$
$$= \frac{\mu_o}{4\pi} I \frac{Rd\theta}{R^2}(\hat{\theta}_o \times \hat{r})$$

Hence, total $\vec{B}$ due to coil at centre O,

$$B = \int_0^{2\pi} d\vec{B} = (\hat{\theta}_o \times \hat{r})\frac{\mu_o}{4\pi}\int_0^{2\pi}\frac{Id\theta}{R}$$

$$= (\hat{\theta}_o \times \hat{r})\frac{\mu_o}{4\pi}\frac{2\pi I}{R}$$

$$= \frac{\mu_o I}{2R}\hat{k}$$

$\hat{\theta}_o \times \hat{r} = \hat{k}$ is a unit vector along z-axis because B is directed along the direction perpendicular to the plane containing $\hat{\theta}_o$ and $\hat{r}$ *i.e.*, xy plane here.

Hence, B is directed out of the plane of the paper, *i.e.*, z-axis.

Example 9:

A current of 10 amp. flows in each of the two conducting wires parallel to each other. The separation between the wires is 2 cm. Find the force per unit length of one of the wires. Will it be a force of attraction or repulsion ?

Solution:

$$F = \frac{\mu_o I_1 I_2}{2\pi r} \text{ Newtons/metre}$$

Here $I_1 = I_2 = 10$ amp. R = 2 cm = 0.02 m,

therefore, $F = \dfrac{(4\pi \times 10^7 \text{ Newtons/amp}^2)\ (10 \text{ amp})\ (10 \text{amp})}{2\pi \times (0.02\,\text{m})}$

$= 10^{-3}$ Newtons/m.

Example 10:

Two similar coils of wire having a radius of 7 cm. and 60 turns have a common axis and are 18 cm apart. Find the strength of the magnetic field at a point midway between them on their common axis, when a current of 0.1 amp. is passed through them.

Solution:

The field is given by equation (1)

$$B = \frac{\mu_0 n\, IR^2}{2(R^2 + x^2)^{3/2}} \text{ Weber/metre}^2$$

The field due to both the coils is given by,

$$B = 2 \times \frac{\mu_0 n\, IR^2}{2(R^2 + x^2)^{3/2}}$$

Here I = 0.1 amp, n = 60, μ_o = 4p $\times 10^{-7}$ weber/amp. metre

R = 7 cm = 0.07 m, x = 9 cm = 0.09 m

therefore, $$B = \frac{(4\pi \times 10^{-7}) \times 60 \times 0.1 \times (0.07)^2}{[(0.07)^2 + (0.09)^2]^{3/2}} \text{ Wb/m}^2$$

$$= 2.5 \times 10^{-5} \text{ weber/metre}^2.$$

Example 11:

If the magnetic field normal to the plane of a circular coil of n turns and radius r which carries a current 1 amp. is measured on the axis of the coil at a small distance h from the centre of the coil, show that it is smaller than at the centre by the fraction

$$\frac{3}{2}\frac{h^2}{r^2}.$$

Solution:

Using equation (1) the field at a distance h from the centre of the circular coil of radius r, is given by,

$$B = \frac{\mu_0 n\, Ir^2}{2(r^2 + h^2)^{3/2}}$$

$$= \frac{\mu_0 n\, Ir^2}{2r^3(1 + h^2/r^2)^{3/2}}$$

$$= \frac{\mu_0 n\, I}{2r}\left(1 + \frac{h^2}{r^2}\right)^{-3/2}$$

Since h is very small as compared with r, therefore, using binomial theorem, we get,

$$B = \frac{\mu_0 n\, I}{2r}\left(1 - \frac{3}{2}\frac{h^2}{r^2}\right)$$

at the centre h = 0.

Therefore, the magnetic field at the centre of the coil,

$$B_o = \frac{\mu_o n I}{2r}$$

The decrease in the magnetic field = $B_o - B$

$$= \frac{\mu_o n I}{2r} - \frac{\mu_o n I}{2r}\left(1 - \frac{3}{2}\frac{h^2}{r^2}\right)$$

$$= \frac{\mu_o n I}{2r} \times \frac{3}{2}\ \frac{h^2}{r^2}.$$

Therefore, the fraction of the decrease of magnetic field.

$$= \frac{B_o - B}{B_o} = \frac{3}{2}\ \frac{h^2}{r^2}$$

Example 12:

An electron has a velocity of 10^4 meters/sec. normal to a magnetic field of 0.1 weber/metre^2. Find the radius of the electron path and also it frequency.

Solution:

From equation (1), the radius,

$$R = \frac{9.1\times10^{-31}\times10^4}{1.6\times10^{-19}\times0.1} = 5.7\times10^{-7} \text{ metre}$$

This is a very small circle. The frequency.

$$n = \frac{v}{2\pi R} = \frac{10^4}{2\pi\times5.7\times10^7}$$

$$= 2.8 \times 10^9 \text{ rPs.}$$

Example 13:

Find the frequency of the oscillator that is to be connected between the dees of a cyclotron with a magnetic field $\vec{B}$ = 1.5 weber/metre^2 for accelerating deutron. How the result is modified if we choose a-particle or proton?

Solution:

$$q = 1.6 \times 10^{-19} \text{ couls.}$$

Mass of deutron = 2 × mass of proton

$$m = 3.34 \times 10^{-27} \text{ kg.}$$

From equation (1),

$$n = \frac{qB}{2\pi m}$$

$$= \frac{1.6\times10^{-19}\times1.5}{2\pi\times3.34\times10^{-27}} = 1.14\times10^{7}\,\text{RPS}$$

$$= 1.4 \quad \text{MC/sec.} \qquad\qquad (1 \text{ MC} = 10^{6} \text{ cycles})$$

Accordingly, the oscillator frequency must be 11.4 MC/sec. If instead of deutron proton is accelerated then,

n = 22.8 MC/sec. since q/m for proton = twice of deutron

and for a = particle n = 11.4 M/C sec. as q/m for a particle = q/m of deutron.

Example 14:

In the above example, if the radius of the dees is 50 cm. find the energy of the emergent deutrons. Also calculate the energy of a = particle or proton if they were accelerated under the same condition.

Solution:

From equation (1),

$$W = \frac{1}{2}\frac{(RqB)^2}{m}$$

$$= \frac{(0.5\times1.6\times10^{-19}\times1.5)^2}{2\times3.34\times10^{-27}}$$

$$= 2.15 \times 10^{-12} \text{ joule}$$

$$= \frac{2.15\times10^{-12}}{1.6\times10^{-19}}\text{eV}$$

$$= 1.33 \times 10^{7} \text{ eV}$$

$$= 1.33 \text{ MeV} \qquad (1 \text{ million electron volt} = 10^{6} \text{ eV})$$

If either a = particles or protons have been used the energy would be doublc, *i.e.*, 26.6 MeV because for a = particle q in the above example will be 2q and m will be 2m.

Hence,

$$W = \frac{4}{2}\left(\frac{1}{2}\,\frac{(RqB)^2}{m}\right) = 26.6\text{MeV}$$

Similarly for proton q s same. But mass is half of the deutron, *i.e.*, q = q, m = m/2,

$$W = 2\left[\frac{1}{2}\,\frac{(RqB)^2}{m}\right] = 26.6\text{MeV}$$

Example 15:

A stream of protons and deutrons in a vacuum chamber enters a uniform magnetic field Both are subjected to the same accelerating potential, hence kinetic energies are the same. If ion stream is perpendicular to the magnetic field and protons move in a circular path of radius 15 cm, find the radius of the path traversed by the deutrons. Given that mass of a deutron is twice that of a proton.

Solution:

The radius is given by equation (1),

$$R = \frac{mv}{qB}$$

So that, radius of the path of proton

$$R_p = \frac{m_p v_p}{qB}$$

and radius of the path or deutron

$$R_d = \frac{m_d v_d}{qB}$$

Charge on deutron is same as on proton and moving in the same field. Thus,

$$\frac{R_d}{R_p} = \frac{m_d v_d}{m_p v_p}$$

Since they have same kinetic energies, *i.e.*,

$$\frac{1}{2} m_d\, v_d^{\,2} = \frac{1}{2}\, m_p\, v_p^{\,2}$$

or

$$\frac{v_d}{v_p} = \sqrt{\frac{m_p}{m_d}}$$

Since $m_d = 2\, m_p$, so

$$\frac{v_d}{v_p} = \sqrt{\frac{m_p}{2m_p}} = \frac{1}{\sqrt{2}}$$

Hence, $$\frac{R_d}{R_p} = \frac{2m_p}{m_p} \cdot \frac{1}{\sqrt{2}} = \sqrt{2}$$

or $R_d = \sqrt{2}R_p$ since $R_p = 15$ cm (given)

hence $R_d = 0.15 \times \sqrt{2} = 0.212$ metre = 21.2 cm.

Example 16:

A cyclotron accelerating protons has dees of 40 cm. radius and a radio frequency supply of 12 megacycles per second at 10000 volts maximum value. Calculate.

(1) The magnetic field B

(2) The speed and kinetic energy acquired by protons

(3) Maximum number of revolutions and time spent by a proton within dees.

Solution:

(1) From equation (1),

Cyclotron frequency $n = \frac{qB}{2\pi m}$

or $$B = \frac{2\pi mn}{q}$$

$$= \frac{2 \times 3.14 \times 1.67 \times 10^{-27} \times 12 \times 10^{6}}{1.6 \times 10^{-19}}$$

$$= 0.79 \text{ weber/metre}^2$$

(2) From equation (2),

Velocity acquired by proton,

$$v = \frac{qBR}{m}$$

$$= \frac{1.6 \times 10^{-19} \times 0.79 \times 0.40}{1.67 \times 10^{-27}}$$

$$= 3 \times 10^7 \text{ metre/sec.}$$

Energy $= 1/2\ mv^2 = 1/2 \times 1.6 \times 10^{-27} \times (3 \times 10^7)^2$ joules

$= 7.5 \times 10^{-13}$ joules

$$= \frac{7.5 \times 10^{-13}}{1.6 \times 10^{-19}} \text{eV}$$

$= 4.7 \times 10^6$ eV.

Number of revolutions inside dees

$$N = \frac{WeV}{2qV} \times 1.6 \times 10^{-19}$$

$$N = \frac{4.7 \times 10^6 \times 1.6 \times 10^{-19}}{2 \times \times 1.6 \times 10^{-19} \times 10{,}000}$$

$= 235$

Time spent within dees = 235/n

$$= \frac{235}{12 \times 10^6}$$

$= 195 \times 10^{-5}$ sec.

Example 17:

A cathode-ray tube with electrostatics deflection has an accelerating voltage, V_a = 1500 volts, length of deflecting plate l = 1 cm. Spacing between the plates (d) = 1 cm and screen is placed at a distance L = 29.5 cm from the end of the deflecting plates. Find the voltage required to deflect the spot by 1 cm on the screen. Assuming that there is no fringing of the field.

Solution:

From eq. (1)

$$\frac{V_d}{Y} = \frac{2d\ V_a}{l(L + l/2)} = \frac{2 \times 10^{-2} \times 1500}{10^{-2} \times (29.5 + 0.5)10^{-2}}$$

or $$\frac{V_d}{Y} = 10{,}000 \text{ volts/metre}$$

or $$\frac{V_d}{Y} = 100 \text{ volts/cm}$$

The required voltage is 100 volts.

Example 18:

A cathode-ray tube with magnetic deflection has an accelerating voltage V = 1500 volts, a magnetic deflecting field is uniform over length l = 2 cm and screen is placed at a distance L = 29 cm from the end of the deflecting field. Find the magnetic field B required to deflect the spot of an electron beam 1 cm on the screen.

Solution:

From Eq. (1),

$$\frac{B}{Y} = \frac{1}{l(L+l/2)}\sqrt{\frac{2m\,V_a}{e}}$$

For an electron on substituting e and m, we get,

$$\frac{B}{Y} = \frac{1}{l(L+l/2)}\;3.38\times10^{6}\sqrt{V_a}$$

or
$$\frac{B}{Y} = \frac{3.38\times10^{6}\times1500^{1/2}}{2\times10^{-2}(29+1)10^{-2}}$$

or
$$\frac{B}{Y} = 2.18 \times 10^{-2} \text{ weber/metre}^2\text{/metre}$$

or
$$\frac{B}{Y} = 2.18 \times 10^{-4} \text{ weber/metre}^2\text{/cm}$$

Therefore the required magnetic field is 2.18×10^{-4} weber/metre^2.

Example 19:

Calculate the value of the electric field intensity which will give a proton an acceleration equal to the acceleration due to gravity.

Solution:

$$\text{Force} = eE = mg$$

or
$$E = mg/e$$

$$E = \frac{1.67\times10^{-27}\times9.8}{1.6\times10^{-19}}\text{volt / cm}$$

$$= 10.2 \times 10^{-19} \text{ volt/cm.}$$

Example 20:

Two parallel plates are kept in air 1 cm apart and potential difference of 100 volt is applied. If there be an electron resting on the plate at lower potential. Calculate (1) the force on the electron, (2) the kinetic energy of this electron, when it reaches the other plate.

Solution:

$$E = \frac{V}{d}\;\frac{100}{0.01} = 10^{4}\ \text{volt / cm.}$$

Force $F = eE = 1.6 \times 10^{-19} \times 10^{4} = 1.6 \times 10^{-15}$ Newton

Acceleration

$$a = \frac{F}{m} = \frac{1.6\times10^{-15}}{9.1\times10^{-31}} = 1.75 \times 10^{15} \text{ metres/sec}^2$$

velocity when it reaches the other plate,

v = Ö2ad Since $v^2 = u^2 + 2aS$ and

$u = 0,\ S = d$

or $$v = \sqrt{2 \times 1.75 \times 10^{15} \times 0.01}$$

$$= 5.9\times10^{6} \text{ m/sec.}$$

Therefore, the kinetic energy

$$= \frac{1}{2}mv^2 = \frac{1}{2}\times9.1\times10^{-31}\times\left(5.9\times10^{6}\right)^2$$

$$= 1.6 \times 10^{17} \text{ joules.}$$

Example 21:

What must be the initial velocity of electrons which can just strike a conductor at – 1200 volts potential.

Solution:

The electrons, having negative charge, are repelled by a conductor at negative potential. In order that an electron just reaches the conductor, it should have the kinetic energy given by,

$$\frac{1}{2}mv^2 = eV$$

$$v = \sqrt{\frac{2eV}{m}} = \sqrt{\frac{2\times1.6\times10^{-19}\times1200}{9.1\times10^{-31}}}$$

$$= 2.06 \times 10^{6} \text{ metre/sec.}$$

Example 22:

A positive ion of charge q and mass m is accelerated by a potential difference V, passes with a constant velocity through the space between two plates separated by distance d. Show that transit time between plates is $\dfrac{d}{(2qV/m)^{1/2}}$.

Solution:

From equation (1),

$$v = \sqrt{\frac{2qV}{m}}$$

If t be the transit line between the plates,

$$t = \frac{d}{v} = \frac{d}{\left(\frac{2qv}{m}\right)^{1/2}}$$

Example 23:

The electric field between the plates of a cathode-ray oscilloscope is 1.2 × 10⁴ Newton/coul. What deflection will an electron experience if it enters at right angles to the field with a kinetic energy of 2000 eV? The plates are 1.5 cm long.

Solution:

Using equation (1), $y = \frac{qE_y}{2m\, v_x^2} x^2$

As given $1/2\ m\ v_x^2 = 2000$ eV

$$y = \frac{qE_y x^2}{4 \times \left(2000 \times 1.6 \times 10^{-19}\right)}$$

$$= \frac{1.6 \times 10^{-19}\,\text{coul} \times 1.2 \times 10^4\ \text{Nt / coul} \times (1.5 \times 10^{-2}\,\text{metre})^2}{4 \times 3.2 \times 10^{-16}\,\text{joul}}$$

$$= 3.4 \times 10^4 \text{ metre}$$

$$= 0.34 \text{ mm}$$

Example 24:

An electron beam is accelerated through a potential difference of 1000 V. An electric field is applied at right angles to the path of the beam by means of rectangular parallel plates. The distance between the plates is 2.5 cm and potential difference 50 V. The edges of the plates are at a distance of 25 cm and 30 cm from the fluorescent screen. What is the displacement of the fluorescent spot ?

Solution:

$$E = \frac{50}{2.5 \times 10^{-2}} \frac{\text{Volts}}{\text{metre}}$$

Kinetic energy = $1/2\ mv^2 = eV$

$$1/2 \times 9.1 \times 10^{-31} \times v^2 = 1000 \times 1.6 \times 10^{-19}$$

On solving this for v, we get,

$$v = 5.6 \times 10^7 \text{ metre/sec}$$

Using equation (1).

$$y = \frac{qE}{m}\left(\frac{l}{2} + L\right)\frac{l}{v^2}$$

We have omitted the subscript and assumed that E is along y-axis and particle is moving along x-axis.

Substituting $L = 25 \times 10^{-2}$ metre

$$l = 5 \times 10^{-2} \text{ metre}$$

and v and E calculated above, we get,

$$y = \frac{50 \times 1.6 \times 10^{-19}}{2.5 \times 10^{-2} \times 9.1 \times 10^{-31}}\left[0.25 + \frac{0.05}{2}\right]\frac{5 \times 10^{-2}}{\left[5.65 \times 10^{7}\right]^2}$$

$$= 0.153 \times 10^{-2} \text{ metre}$$

$$= 0.153 \text{ cm.}$$

EXERCISES

1. Obtain an expression for the magnetic field along the axis of circular coil carrying current.
2. Give the theory and working of Helmholtz galvanometer. How do you get uniform magnetic field between the coils?
3. A closely wound solenoid of 1200 turns has an axial length 80 cm and radius of 1.5 cm. A current of 1.2 amp flows in the solenoid. Find the flux density at the middle of the axis.
4. A long solenoid of length 1 metre and radius of cross-section 1.5 cm has five layers of windings of 850 turns each. If the solenoid carries a current 5.0 amp. calculate the values of

   (1) Magnetic induction and

   (2) Magnetic flue ϕ_B for a cross-section of the solenoid at the centre of the solenoid.
5. What are vector and scalar magnetic potentials? How the idea of vector potential is introduced? Obtain an expression for magnetic vector potential due to an electric current flowing in a wire and hence calculate the magnetic field.

6. Discuss the equivalence of magnetic shell and circuit carrying electric current.

7. Two long parallel wires, each carrying a current of 10 amp. in the same direction are separated by 10 cm. Calculate the force between the wires per unit length.

8. A square loop of wire of edge 10 cm, carries a current of 10 amp. Calculate the value of B at the centre of the loop.

9. Two long straight wires are kept in free space at a distance of 1.0 metre and 1.0 amp current is flowing in the same direction in each wire. Calculate the direction and magnitude of the mutual force per metre between the two wires. What will be the difference if the current flows in mutually opposite direction?

10. A current of 30 amps flows in a long wire, a rectangular loop ABCD carrying current 20 amp is situated 1.0 cm away from this wire. If the side of loop facing wire is 30 cm and other side is 8 cm. Calculate the magnitude and direction of the resultant force acting on the loop due to magnetic field of the current in the straight wire.

11. The axis of circular coil of radius 5 cm. having 10 turns and carrying a current of 5 amp. makes an angle q with a uniform magnetic field $\vec{B}$. Find the torque on the coil.

12. A current of 30 amp. flows in a flat circular coil having 100 closely wound turns of radius 10 cm. What is flux density at the centre of the coil?

13. What is Biot-Savart law? Apply it to determine the magnetic field.

    (1) Due to a long straight wire carrying current.

    (2) Due to a wire of finite length carrying current.